Exploring The Route of Parks
A journey to 17 National Parks
SOUTH AMERICA
Alerce Andino
Puerto Montt
Hornopirén
Pumalin Douglas Tompkins
Chaiten
Corcovado
Melimoyu
Queulat
Isla Magdalena
Coyhaique
Cerro Castillo
Patagonia
Laguna San Rafael
Cochrane
Tortel
CHILE
ARGENTINA
Patagonia Region
Bernardo O'Higgins
South Pacific Ocean
South Atlantic Ocean
Torres del Paine
Puerto Natales
Pali Aike
Kawésqar
Punta Arenas
Tierra del Fuego
Yendegaia
Ushuaia
Alberto de Agostini
Cabo de Hornos
N
W
E
S
National Park
route by car
route by ferry
200 miles
© emk.nl

A WILD IDEA

Also by Jonathan Franklin

438 Days: An Extraordinary True Story of Survival at Sea

33 Men: Inside the Miraculous Survival and Dramatic Rescue of the Chilean Miners

A WILD IDEA

Jonathan Franklin

HarperOne
An Imprint of HarperCollins*Publishers*

HarperCollins books may be purchased for educational, business, or sales promotional use. For information, please email the Special Markets Department at SPsales@harpercollins.com.

FIRST EDITION

Designed by THE COSMIC LION
Mountain art by Nur Hasan Icon/Shutterstock
All maps © emk.nl; used by permission.

Library of Congress Cataloging-in-Publication Data is available upon request.

ISBN 978-0-06-296412-0

21 22 23 24 25 LSC 10 9 8 7 6 5 4 3 2 1

Totinha—I adore you!

You've made these past
four years the best of our life.

He was just not impeded by any kind of realities or practicalities—it was like, What do you want? Don't limit your potential with a lack of imagination. If you don't think big, it's never going to happen. I don't think he cared too much what other people thought. I don't think he was trying to please anybody else. And "get out of my way, I'm going for this vision that I've created." And that's potent. It's heady.

—**QUINCEY TOMPKINS**, oldest daughter of Doug Tompkins

He was fearless. He never turned his back on a difficult problem, a difficult conclusion, or an overwhelming challenge—an overhanging cliff on a climb, or a planet where loss of habitat and global warming were putting the future in danger. Doug never stopped, never slowed; he was the quintessential man of action—a force of nature, for nature.

—**LITO TEJADA-FLORES**, filmmaker and climber

At the summit, we got nailed by this huge ice storm, and Doug thought he knew the way down and started leading us. Suddenly he stopped. He froze. He was at the edge of a precipice a thousand feet down. He was completely wrong. Then I pulled a compass out—what I call a Boy Scout compass. We were 180 degrees off and Yvon Chouinard gets right up in my face and said, "Isn't this great! Makes the whole trip worth it!" And I am thinking, What kind of maniacs am I with?

—**TOM BROKAW**, NBC television anchorman,
author of *The Greatest Generation*

Doug learned fast; he felt like he was invincible because the normal amounts of risk didn't apply to him. If you ever flew with Doug in his plane, you'd know what I mean. He took us on a flight over Patagonia, and it was all I could do to keep my stomach down. He would set it on one wing and spin around and around looking at something. He clearly had that quick-twitch, fighter-pilot stuff.

—**DAVE SHORE**, rafting guide

Contents

Author's Note

I lived many years in awe of Doug Tompkins. He'd done the impossible—climbed to the heights of corporate America and then cashed in his chips, taken his millions, and fought for nature. I admired his fight for native forests, grasslands, rivers, and wetlands. I visited many of the remarkable national parks he created in South America. When my Uruguayan friend Rafa called to tell me that Doug had died in a kayak accident in December 2015, I immediately felt like an idiot. How had I not spent more time with this genius in my midst? How had I only interviewed him a half-dozen times in his last ten years?

As I set off to write this book I sought to capture the essence of his remarkable life, and I felt qualified for the challenge. My passion for the great outdoors set in early—bouldering in New Hampshire and swamp exploring in Massachusetts. Like Doug, I was a downhill ski racer and loved testing the limits of speed and control. Doug took his passions to San Francisco, then in 1989 flew down to southern Chile. I too followed a similar path, traveling from San Francisco in 1989 to explore southern Chile by mountain bike. Like Doug, I have spent half my life in the United States and half in South America.

When I first queried Kris Tompkins for permission to write a book about her late husband Doug, I knew I didn't have to ask. In fact, she confirmed that when she told me, "You don't need my okay; you can just go and write it." Kris was right, but that was never the book I wanted to create. I was interested in the inside story of a remarkable man. I interviewed Doug enough during my nineteen years reporting for *The Guardian* to know his close friends were protective and almost tribal in their secrecy. Then Kris said no. They couldn't help me on the book. Eight months later, I asked Kris again for her permission and collaboration. Again I was told no. Kris was too busy.

I tried again. If she didn't have the time, could she at least give her blessing? Could we agree to walk separately down the same path? The biography would not be authorized but collaborative. We could share notes, impressions, and stories in the wake of her husband's shocking death. Here we found a common ground that, after two years, blossomed into a collaboration beyond my imagination and eventually hours of face-to-face conversations with Kris.

Kris received my calls, read my emails, gave long interviews in Valle Chacabuco and Pumalín Park in Chile, met with me in Rincon del Socorro in Argentina, and, as I was finishing the book, made time for video calls from her home in California. She shared Doug's love notes, private emails, personal photo collection, and stories of their life together. She shared his passion for conservation and was generous in sharing his life with me.

For nearly four years, I explored the world of Doug Tompkins. At his park in Patagonia I took an afternoon walk through Chacabuco Valley, aware that wild pumas roamed nearby. Instead of scary it was invigorating; a sense of longing awoke inside me—as if I needed a reminder that humans too are part of the food chain.

While visiting the national parks Doug had established in Argentina, I could hear but not see a family of howler monkeys high in the trees. My seven-year-old daughter, Akira, mimicked the call and the howlers descended, curious about this strange little monkey—roughly their size. For minutes they stared eye to eye at one another, calling back and forth. Akira continued mimicking monkey talk, and the howlers chattered among themselves excitedly as they discussed the meaning of this new voice in the woods. Everyone present felt the connection, the communicating, the communion. I had brought my children on the reporting trip hoping that in the wild spots that Doug Tompkins saved perhaps some wild would rub off. Might they return home themselves rewilded? Akira's encounter with the howlers made it all seem possible.

This book is my journey into one man's love affair with the wild. The wild mountains, forests, and rivers. Doug Tompkins was also wild. Competitive, hyperactive, and as flawed as his friend and neighbor Steve Jobs, with whom he argued at parties. Tompkins was an environmentalist who drove a red Ferrari. A multimillionaire who preferred to sleep on a friend's couch. He was a stickler for details, yet he hardly noticed his two daughters right before his eyes. Hard-nosed, arrogant, and argumentative, he despised compromise.

The world for Doug Tompkins was black and green: you were either a scourge or a seedling. He never seemed to worry what others thought. When the media savaged him, Tompkins laughed. He told Thomas Kimber, a young entrepreneur who was his neighbor, "It doesn't matter; in fifty years they will be building statues of me."

Like the way he drove cars and paddled kayaks, Doug Tompkins rarely looked back. Yet, for all his faults, I found myself transfixed by this rock climber who at the age of forty-nine and atop the peak of capitalism took a deep look around and admitted to himself, "I've climbed the wrong mountain."

To understand this complex man, I conducted roughly 165 interviews ranging from his seventh-grade classmate Stone Ermentrout to his lifelong friend Yvon Chouinard. I sat for interviews with Susie, his first wife; both his daughters; and dozens of employees who loved him plus a half dozen who loathed him. Notably, by the time of his death, Tompkins had converted many adversaries into allies. Critics who thought he was exaggerating the demise of planet Earth—people who wondered what he meant when he said that extinction was the "mother of all crisis"—began to see that he had offered them not a doomsday scenario but a glimpse of the future.

The planet has never needed defenders more than now. Wherever we look, environmental news is bad news. Forest fires. Global warming. Extinction. It's a litany of loss and destruction that Doug Tompkins fought so hard to slow. He liked to quote his mentor Arne Naess and say he was "a pessimist for the twenty-first century and an optimist for the twenty-second century." Despite his pessimism about human behavior, Tompkins maintained faith that the Earth could recover.

During the COVID lockdowns it became evident to many urban and suburban residents that the animals are still out there. Pumas have visited silent urban centers, turtles have hatched on untouristed beaches, and dolphins have explored suddenly quiet coastal waterways for the first time in decades. Given a respite, a chance to breathe, nature is resilient. But change requires action. Tompkins quoted the naturalist author Edward Abbey, who mocked the idea that economic growth was a measure of economic health. "Growth for growth's sake," Abbey once cracked, "is the philosophy of the cancer cell."

For Tompkins, the key to environmental health was not growth but stability. He recognized that on a finite planet there is a deep need to give as much as we take and that before we die, we must leave the world a little bit better.

As I wrote about the life of Doug Tompkins, I attempted to describe this book in a narrative that my eleven-year-old daughter, Zoe, could understand. I sought to help her, even at that young age, understand the concept of a legacy with dignity. I told Zoe that Doug Tompkins—whom she'd heard quite a bit about during her dad's four-year journey—felt that a most noble goal in life was to leave the planet "a little bit better." She smiled, nodded, and then asked the kind of question so natural to a child and so poignant for an adult: "Why only a *little* bit better?"

Jonathan Franklin
Punta de Lobos, Chile

Part I

Chapter 1

The Rucksack Revolution

Doug is the kind of person you could throw out in a desert naked with a stick, and within a couple of weeks he'd have an empire. I think he's the most street-savvy human being I've ever known. Wherever he landed, whatever situation he was in, he worked out all the street angles way faster than anyone else. It was not always according to the book, or legal, *but it worked.*

—DICK DORWORTH, the world's fastest skier in the 1960s, a frequent travel companion of Doug Tompkins's

Doug Tompkins paced the San Francisco sidewalk, hawking downhill skis to beatniks and groggy sailors. Fast on his feet and quick with a retort, the twenty-two-year-old thrived on the weave and bob of the street salesman. The charismatic high school dropout badgered customers to peruse his eclectic collection of custom-forged climbing pitons and fishermen's sweaters from Scotland that he promised were resistant to any winds. Like a fencing champion, he parried back and forth with the pedestrians outside his store, trying to sell, sell, sell. "Need a sleeping bag? Wool pants? An ice ax?" he crowed to onlookers ambling past his quaint storefront in North Beach.

It was 1965. The store had been open barely a year and cash was tight at the tiny startup. Doug's entire $5,000 budget was gone—spent on a storewide overhaul and a meager pile of equipment. Salaries were reduced by convincing climbing buddies to help with the street hustle. Compensation included a front-row seat at the hottest street scene in all of San Francisco, plus free beer.

Tompkins loved to dress up. Like a circus performer, he often switched outfits, and whether it was with his black top hat or his furry, ankle-length jacket, he appeared stylish yet brash. So too his shop. When customers asked why he named his retail outlet "The North Face," Tompkins answered with the brazen air of a man who loved bombing down a mountainside on wooden skis and reaching speeds approaching ninety miles an hour. "The south face is the most often climbed, the snow is softer, and the sunlight makes it warmer," he said with a sigh. "I prefer the more difficult side. The hard, icy face. The North Face is a more difficult challenge. I take that route in life."

Although it was decades away from being a globally recognized clothing company, the original store called The North Face, created by Doug Tompkins as a twenty-one-year-old, was just the first of his three mighty brands that he would turn into world-renowned enterprises over the course of his life. Disruption was his game and, according to his best friend, Yvon Chouinard, founder of Patagonia clothing company, Doug liked nothing better than breaking the rules. "If you want to understand the entrepreneur, study the juvenile delinquent. The delinquent is saying with his actions—*This sucks! I'm going to do my own thing!* That was Doug. That's why he dropped out of school, because, like a lot of kids, he couldn't tolerate just sitting at a desk and being told what to do, and he just had too much energy for that and too many ideas floating around in his head."

In front of the plate glass display windows featuring an oversized Bob Dylan poster, Tompkins and Chouinard harangued pedestrians walking through the North Beach neighborhood—some on their way to Chinatown, others coming up the hill from the docks. Yvon plopped Quincey

(Doug's infant daughter) atop his shoulders as a conversation starter to lure customers in. When Quincey tired, he lay her down for a nap in the display window. The cute baby, naked as the day she was born, sleeping atop a pile of fluffy reindeer-skin rugs became a hit in the growing lore of the community.

Besides selling newfangled skis and cutting-edge mountaineering equipment and clothing, Doug's store was becoming known as the place where "the baby sleeps on the reindeer rugs," and inside the shop there were often raucous gatherings, intense debate, and an eclectic crowd feeding off what was quickly becoming "the North Face scene."

Tompkins and Chouinard bickered and bantered like a comedy team as they pestered pedestrians. They kept up a running commentary on the ridiculous nature of their proposition. Who sold 120-foot, spun-nylon Austrian climbing ropes to half-drunk sailors looking to party? To their right was Big Al's Saloon. To their left, dancers enticed businessmen into The Condor Club—a novel strip bar where in the evening, when the music cranked up and a white grand piano was lowered down from the ceiling with the busty Carol Doda coyly splayed on top, the floorboards at The North Face reverberated with tremors, like a small earthquake. Between shows, the strippers hung out with the climbers. Tribe to tribe they shared the thrill of breaking traditions and routines. "He looked totally out of place in that shop. The strip-show barkers outside their bars, a horde of low-lifers cruising Broadway," said Chris Jones, a fellow climber who met Doug when he purchased downhill skis. "But he was always overflowing with energy."

Boxed between dance revues and a string of sailor bars, The North Face store was center stage for the North Beach neighborhood's explosive cultural upheaval. Across the street, the poet Lawrence Ferlinghetti ran City Lights bookstore, and provoked the nation's leading moralists by publishing genre-breaking poetry and fiction. Allen Ginsberg's poem "Howl" incensed the censors, who frothed at its celebratory references to illicit drugs and roaringly good sex. The uproar, of course, just made "Howl" that much more delicious for those with an appetite for rebellion.

As the carnage from the Vietnam War cleaved apart the national psyche, Doug was at street level taking it all in. "Drugs were just one part of a whole cultural, social circle at that time—drugs, music, permissive sex," Tompkins later wrote. "It was a whole change of value systems going on then and a tremendous experimentation. Of course it was a very broad swing of the social pendulum. A new morality was being formulated. And I was right in the eye of the social revolution."

An Italian immigrant neighborhood in the '50s, North Beach in the '60s was not as counterculture as Haight-Ashbury several miles to the west. Unlike "The Haight," with its flood of youth who were seeking refugee status from mainstream America, North Beach was more like the staid '50s than the revolutionary '60s. But change was in the air, and in North Beach it took the form of tourists off the beaten track and passing revolutionary poets. Across the Bay, the Berkeley Free Speech movement at University of California's Sproul Hall in Berkeley and Mario Savio's passionate call on that campus to "put your body upon the gears" lit a fire on college campuses nationwide. In Oakland, the Black Panther Party fought for the cause of black liberation. Across the nation, J. Edgar Hoover's FBI agents worked overtime to sabotage, squelch, and illegally upend the incipient uprisings.

At street level in North Beach, Doug and Yvon saw it all. They faced a daily ebb and flow of beatniks, tourists, and drunks. They loved the buzz. Doug's wife, Susie Russell, added her own touch. She bought bikinis in France and then hawked them to her posh friends.

When they needed a rest from the hustle, Tompkins and Chouinard walked to the back of the street-level store and descended a staircase with missing planks, past a wall of exposed electrical wiring. In the basement they grabbed a snack and rested. The dirt-floored room was cool and dank. It was lit by a naked bulb and built into the side of the hill with an uneven floor that made the entire basement feel crooked, even when they didn't smoke a joint.

In the basement Chouinard stored boxes of climbing equipment that he hammered out at his beachside forge in Ventura, California. A mechanical whiz and precise builder, Chouinard banged out rock-climbing gear and when he went to Yosemite, the trunk of his car was stuffed with pitons and climbing gear. The tribe of climbers could be counted on to spend their few dog-eared dollars on a reliable tool from Chouinard Equipment. Doug helped his buddy with wholesale orders, as Yvon was stuck with a three-year US Army tour of duty and couldn't manage the business while deployed in Korea. Yvon was finally getting out of the army but was stuck in limbo and spent his days killing time working cleanup duty at the nearby Presidio Army Base in San Francisco.

Although he had little cash for inventory and no cash for marketing, Doug anticipated the rising popularity of outdoor travel and adventure, so he set about stocking his small store with trendsetting designs and innovative gear. A trickle of repeat customers passed the word—The North Face was *special*. "The big hits were the reindeer skin rugs, which were a novelty item and flew out like hotcakes," said Duncan Dwelle, the first store manager at The North Face. "Then, the striped fishermen's sweaters. And the bikinis. They were completely unrelated! They were illustrative of Doug's vision of merchandising. His vision was that we were trying to present a style. But he couldn't articulate it. I don't think he really understood it at all, then. But what he was selling was a sense of style. I was much more explicit about it—regarding mountaineering and backpacking. The style was: if you go light, you'll have more fun."

As accomplished climbers and explorers, both Doug and Yvon mocked equipment junkies. They gagged at the proliferation of nonessential gadgets. According to their credo, anything that wasn't practical was dead weight. They were not dilettantes; they were obsessed climbers: gritty, unshaven members of a tribe known affectionally as "dirtbags." They were hardcore outdoor enthusiasts who whittled down the handle of a toothbrush to lighten the load.

Despite its unusual location sandwiched between bars and strip clubs, The North Face shop drew crowds. The walls were adorned with dramatic photographs of downhill ski racers. The wooden signs, carved by hand, suggested a personal touch. Customers strolled across a woolen green carpet and past skis set against the whorled grain of redwood shelving. A display shelf arranged and labeled Yvon's hand-forged pitons and carabiners as if they were artifacts from a long-lost culture. Sleeping bags dangled from the ceiling. Pedestrians popped in to gawk. This place had an *aura*, and many visitors studied the product display cases as if they were strolling through a museum. The poet Allen Ginsberg was a regular visitor. He lived in the neighborhood, hung out at the City Lights bookstore, and felt taken by the beauty of The North Face store. He seemed enchanted by an oversized Ansel Adams photograph of Yosemite National Park. "Ginsberg stared and stared at the pictures of El Capitan," said Lito Tejada-Flores, a ski instructor who worked part time for his friend Doug at The North Face. Said Lito, "Ginsberg never understood how anyone could dare scale that."

"We were all sort of social rejects, Chouinard and these guys. We were the guys who didn't really want to make it in high school," said one Yosemite climber, Joe McKeown. "I didn't play football or wrestle. I'd come back from a summer of climbing in Yosemite in the early '60s, and the football coach would come over and say, 'What's the matter, McKeown, you yellow son of a bitch—aren't you going to try out for the football team?'"

"We were more climbers than we were '60s hippies. There was interesting stuff happening. But that wasn't us. We were on our own trip," said Lito. "Climbing is an obsessive passion. It's more than a sport. Tennis is a sport, but when you're a climber, you're committed to it because it's exciting and meaningful and thrilling."

Yvon, Doug, and Lito spent days at a time in Yosemite Valley. They slept at Camp 4, a rambunctious campsite for climbers and a perpetual headache for National Park Service administrators. Climbing by day, stealing away from park rangers by night, they felt like outlaws and called themselves "The Valley Cong"—an ideological tip of the hat to peasants resisting the US invasion of Vietnam and the carpet bombing that killed more than two million civilians. "Climbing is the exhilaration of good health," Tompkins told an interviewer. "You have to go out and exercise and get the body going to do those things to feel the exhilaration of being healthy. The blood is going around and when you get that feeling, it's reinforcing and you can do it again. You get into a cycle."

Although he was far from being a full-time member, Tompkins earned respect from the Camp 4 tribe on a four-day ascent of Yosemite's Salathé Wall where he followed Yvon's lead. It was only the fourth time that the 3,000-foot wall had been climbed.

Inside Yosemite Valley, as he navigated smooth granite slabs, Doug felt energized. The life-threatening, edgy challenges of rock climbing melded his exceptional coordination and balance with a love for spontaneity. Real-time problem solving and finding a first-ever route to a summit was exhilarating. "Doug and Yvon used to talk about walking all two hundred and ten miles of the John Muir trail in an overcoat with only the food they could carry in their pockets," said Dwelle, The North Face store manager, who was often stuck back in North Beach filling out orders and inventory to keep the business running.

Yvon and Doug traveled extensively together. They climbed in British Columbia, the Swiss Alps, and the Scottish Cairngorms. Perpetually short of money, they hitchhiked and camped; rarely did they have funds to pay for lodging. Yvon was jailed for seventeen days in Winslow, Texas, when local police yanked him off a passing freight train and found he had no job, no cash, and no firm plans. "We used to sleep in the Goodwill donation boxes," Yvon said with a laugh. "It was warm with all those clothes, but you'd get woke up whenever people dumped a fresh box of used clothes on top of you."

"Yvon is a quiet, reflective, introverted person," said Dwelle. "Doug is exactly the opposite—a natural-born leader, one in a billion. Extremely charismatic, energetic, full of imagination and ideas. He didn't believe rules pertained to him. Once when he was out biking with several friends, he ran a red light, and a cop chased him down and said, 'Why didn't you stop?' Doug said, 'I thought I could outrun you.'"

Douglas Rainsford Tompkins was christened and bred to be anything *but* a rock climbing dirtbag. Born to descendants from the Mayflower and raised in New York City's Greenwich Village, as a small child his world shone with fine art and antiques. His father, John "Jack" Tompkins, was a glider pilot during World War II who owned a high-end antique furniture business. Jack traveled the United States in his small airplane, and as a youngster Doug sat atop a cushion in the copilot's seat. He could decipher navigation maps while still in middle school. Steering for minutes at a time, Doug flew with his father cross-country as they examined, evaluated, and purchased museum-quality wooden furniture.

Shadowing his father, Doug developed an eye for discovering hidden treasures. In a New England church, Jack inspected a prized refectory table while the clergyman bantered about the value of the piece. Jack just shook his head, no. He looked at the priest and declared, "You don't know how much this table is worth and I do. And what you're asking for that table is not *close* to what it's worth. I've got to pay you more. Here's how much I will be paying you for this table, and I'm still going to do fine." The stunned clergyman accepted a sum far higher than his asking price. As father and son drove back to New York, the elder Tompkins lectured his son that "the only really good business deals are when it works out well for both sides."

With a fine eye for the masterpiece in a collection, Jack Tompkins earned enough money to move the family into farmlands north of Manhattan. They settled in Millbrook, New York, where the neighborhood included Harvard professor Timothy Leary and his infamous LSD crash pad. Rambunctious and entrepreneurial, young Doug was a handful. Faith McCellan, his mother, once tied him to a tree at the beach to keep him from escaping while the family was enjoying the seaside. Doug showed a knack for inventing projects—not only did he love to raise animals but by the age of eight he was converting his flock of chickens into his own business selling eggs. Apart from his brood of chickens, he also raised sheep with such care and attention to details that he regularly snatched up the blue ribbons at the county fair. Business was in his blood. So was a deep competitive streak. On the athletic fields he excelled as a star athlete.

The Tompkins farm housed geese, horses, goats, and rabbits. Set back from the road and surrounded by acres of pastures and pine forests, the New England, country-style home was furnished with antiques that guests were encouraged to pick through—and perhaps even purchase. Family friends and trusted insiders understood that the sequence of numbers and letters on the bottom of each item, a bit like Roman numerals, were actually coded price tags. During dinner parties and after cocktails, Jack and Faith in effect hawked the furnishings in their home.

Doug's father was a Calvinistic taskmaster. Praise from him was sparse. Yet Jack was proud of his genteel living. In a pink-red riding jacket, he assumed the image of an English lord out on a fox hunt. His private airplane and long driveway epitomized his tendency to embrace the illusion of boundless lands. "Never buy a house where you can see neighbors," he cautioned young Doug without understanding how that lesson would shape his son's destiny.

Jack taught Doug the secrets to evaluating antiques. Jack regularly cited a book that illustrated examples of furniture and used photographs to show how every antique could be divided into one of just three categories: the good, the bad, and the exceptional. He made Doug study it. Doug devoured the book and, at a young age, memorized large parts of it, which honed a keen eye for composition and beauty, an aesthetic taste that would sculpt the contours of his life.

The lucrative family antique business reaped benefits, including upgrades to the family automobile and family airplane with enough money left over to send nine-year-old Doug to Indian Mountain, a boarding school in nearby Connecticut that offered a renaissance education along the lines of the ideals of Ancient Greece. At school, Doug read French, wrote Latin, analyzed newspapers, and debated current affairs. At his eighth-grade graduation, he earned honors as the top athlete at the sports-heavy academy.

For high school, his parents enrolled Doug at Pomfret, a Connecticut boarding school with a well-trodden path into Ivy League universities. Doug's aristocratic grooming was on schedule until a rock climbing trip upset his parents' carefully laid plans. "Somebody took me rock climbing as a kid," he explained. "It was a woman, actually. She was having a marital affair with a ski instructor over in the Catskill Mountains. She took me along as a decoy," he added. "She said she was taking me rock climbing, but she was really going to see this guy. He taught me how to climb. Since I was there, he had to! And then I fell in with all the climbers there. They were socially progressive, from the left side of the political spectrum. In some ways they were early 'greens.' Climbing, of course, got me out in the wild outdoors and connected me with the nature tradition. That sent me in another direction. That changed the course of my life completely."

Climbing also introduced him to Yvon Chouinard, whom he met while he was in high school and climbing in the Shawangunks, in upstate New York. Tompkins fell in with a group of rebel climbers who became addicted to life in the "Gunks." They called themselves The Vulgarian Mountain Club in homage to their affinity for disrupting the local social norms.

Once hooked, Doug spent every possible weekend rock climbing on ever more difficult routes. He learned to fix ropes, rappel down a cliff, and hammer pitons into the rock face. In high school Tompkins skipped school on Mondays. Three-day weekends became standard as he stole an extra day to climb or ski—and took another step toward expulsion.

On the ski slopes, he was a skilled daredevil. Schussing straight down the mountainside or weaving through the gates, whether in downhill or slalom, Doug won race after race. He told classmates he was headed to the 1964 Winter Olympics. School, he declared, slowed him down. Why attend classes if he could rock climb and downhill ski? Doug was infamous for being the only one in his class to drive cross-country (without a driver's license) as he escaped with older friends to climb at Devil's Towers in South Dakota. He was not yet sixteen years old.

Just weeks before his high school graduation, Doug was expelled. Rumors suggested that he had pawned antique furniture belonging to the school and kept the cash. His parents were stunned. So was Tompkins. He was free.

"My parents wanted me to go to college," he said. "But by that time, it was too late. I was ready to see the world rather than sit in a classroom. I just quit school. I took high-paying jobs, what we called 'booming.' I was a *boomer*. I worked hard for a short period of time on logging crews as a feller. I saved money and then I took off."

Tompkins spent three summers chopping down huge trees. He strapped on a leather contraption to each lower leg and, using climbing spikes, jabbed his way up the trunk until he was about a hundred feet high. Tompkins then sawed away branches and cut off the treetop. With ropes and saws flapping about, he bounced from one perch to another, delighted with the opportunity for aerial gymnastics. Once the trunk was cleared of branches Tompkins felled the old-growth tree into the underbrush. He excelled at tree felling. In many ways the challenge was an echo of the skills he used in rock climbing. Felling provided a bird's-eye view of the forests—then called "virgin timber"—and put wads of cash in his pocket. When his supervisors noted his agility in the trees, they hired him to plant explosives. At times they needed to place dynamite charges to blow away rock formations and build roads. They learned that Doug's fearlessness allowed him to take on the dynamite missions with gusto. "Dad and I thought we'd starve him out," confessed Faith, his mother. "We said, 'Okay, do what you have to do, but we won't pay one cent. If you go to college, we pay everything.' It didn't work! He never set one foot in college."

Baling hay in Montana for a season, Doug learned to heave the forty-pound loads around, and added a sheath of muscles to his arms and chest. Perpetually tanned and grizzled, Doug cut a dashing figure as he changed from lumberjack garb to party clothes. Any job that could bring him closer to the outdoors was worth taking. He worked as a janitor at the famed Jerome Hotel when Aspen was still a wild outpost in Colorado, and he gained access to the slopes and the winter ski racing circuit ranging from the Rockies to California's Squaw Valley. "He was a really good ski racer. When he set his mind to doing something, he was focused on it," said Billy Kidd, the Olympic skier who raced with Tompkins during the late '50s and early '60s. "In the summers, we went to Portillo in Chile [where it was winter] and trained for months," said Kidd, who shared carousing motorcycle rides in the Chilean Andes with Tompkins just before winning the world championship and becoming the first American to win an Olympic medal in downhill skiing.

During a break in their training, Tompkins and Kidd convinced a local BMW outlet to lend them two motorcycles. They blasted around town until they had an accident and scrambled to hide the evidence. "They spent all night cleaning the bike up and using shoe polish to cover the patches," said Tom Brokaw, the journalist who loved to travel with Doug. "They wheeled the bikes into the dealership and took off. The next day there was an All Point Bulletin out for them. Doug knew [an influential] family in Chile and they got him out of it."

After completing ski training in Chile, Doug hitchhiked by airplane throughout South America. Instead of flying a commercial flight home, the teenage high school dropout staked out South America's small municipal airports and chatted up their air traffic controllers. When a small plane was less than full, Tompkins sold himself as an experienced navigator. With that trick, he flew free on flights from Chile to Peru and into Colombia as he pinballed his way home. In the Amazon jungle near Iquitos, Peru, he worked for a research team that valued his climbing skills. They sent him into the trees to catch monkeys.

"If he really had set his mind on making it to the Olympics as a ski racer, then it would take something like a knee injury to keep him from accomplishing that goal," said Kidd, who knew Tompkins from competitions in New England. "And, unfortunately, in ski racing, you have a lot of injuries, and especially in those days when we had wooden skis and leather boots, and the release factor in the bindings was only when the screws ripped out of the ski."

Claude Suhl, a fellow climber, described Tompkins as once suffering an injury that slowed his path and knocked him out of the elite competition. At the time, Tompkins could still ski, but not at the level that provided a shot at the 1964 Olympics. He was not invited to join the US downhill ski team. Fueled by his fierce athletic talents and an obsession to win, Tompkins focused on rock climbing and founded California Mountain Guide Service (CMGS), a prelude to outdoor education programs like NOLS, the National Outdoor Leaderships School, formed three years later.

The instructors at CMGS were highly regarded climbers, including Chuck Pratt, Tom Frost, Royal Robbins, and best buddy Yvon Chouinard. To market CMGS, Tompkins designed a catalog featuring photographs of the mountains they would climb. Despite the elegant brochure, his clients were classic dirtbags that one friend described as "little more than high school students packed into a battered van." As Tompkins promoted his climbing school, so many friends and clients peppered him with questions about equipment that he decided to open a small business. From inside a garage in Berkeley, Doug sold camping and climbing equipment. "He couldn't stand still," said Yvon. "He was very entrepreneurial and was always coming up with ideas. I don't know if he was a great businessman, but he was willing to take risks and try new things. He doesn't take advice too easily and he never liked anybody telling him what to do and hated authority, so he was pretty much going solo on a lot of this stuff."

In the fall of 1962, Tompkins walked out of the forest near Emerald Bay State Park, on the shores of Lake Tahoe in California's Sierra Nevada mountains. He was ruggedly handsome, tightly muscled, totally broke, and in need of a ride to San Francisco. He started hitchhiking. A car pulled over. Tompkins climbed in and the driver, Susie Russell, thought the hitchhiker looked like a handsome bandit, with his rough clothes and coils of climbing ropes strapped like bandoliers across his chest. As he sat down, Susie asked, "Where are you from?" Not missing a beat, Doug swatted the question back with a cocky "Back east. The better coast."

Susie, who was an independent nineteen-year-old and earning hundreds of dollars as a Keno runner in a Reno casino, didn't think twice as she dropped Doug in San Francisco, telling herself she was good to be rid of this "arrogant lumberjack." A week later, at work, she received a message. Doug, *that guy she had picked up hitchhiking*, was in a jam. He was in jail for shoplifting a steak from a Safeway supermarket. Could she post bail? Susie lent him all the $65 cash from her cash pouch and never expected to hear from Doug again. But he repaid the loan and gifted her with flowers liberated from the dumpster at the funeral parlor next door to his San Francisco apartment. Doug soon invited Susie to dinner. It went well, and they made plans to escape to Mexico in a Volkswagen van. "He picked me up at my mother's house," said Susie. "I had a pair of shorts, a T-shirt, a bikini, and no shoes. And off I went."

Within a year they were married at the Tompkins home in Millbrook, New York. After the wedding, they packed a few pieces from Jack's antique business in the back of their beat-up van and headed west, first to deliver the furniture, then to settle in Northern California. The van was missing windows, so they drove and rode while sitting inside sleeping bags. It was 1963 and they were carefree, nomadic, and early arrivals to the revolution. Rumblings of the hippie, psychedelic, and political uprising that commenced in California were about to spread throughout the United States.

Settling at a lodge in Squaw Valley, California, Doug worked on the ski patrol and Susie waitressed. They lived in a rented basement, trying to scrounge up money from the sale of "Doug's Rugs"—one of his many efforts to launch a brand, in this case finely woven Persian rugs. "Doug had a big drive and a lot of energy, and he was very physically oriented," said Susie. "But I think that a lot of that came from the fact that his father was so unfair to him and cruel to him. He never complimented him. Never acknowledged him. It was really hard." Sensitive to his lack of academic credentials, Tompkins carried a chip on his shoulder the size of the high school diploma he never obtained. "He told people he was a Yale student," said Susie. "He was creating this person that he wanted to be and was setting the stage."

As his business selling ski gear and climbing supplies expanded, Tompkins moved his tiny shop from the garage in Berkeley to a less expensive basement location in San Francisco. In spring 1965, the owners of the Swiss Ski Shop, whose basement Doug had rented for his store, shut down for the summer, giving Doug a brief run of the street-level space, as well. Although he had only a ninety-day sublet, he launched a complete renovation of the entire store. Exploring north of San Francisco in rural Sonoma County, Doug discovered abandoned chicken coops framed with weathered redwood planking. Looking at the grain, admiring the hues, he was reminded of the antique furniture his father coveted and valued. Tompkins envisioned the redwood planks as wall paneling for his retail space. The chicken farmers viewed the half-collapsed structures as a fire hazard and eyesore, so they thanked the energetic young man from San Francisco who didn't charge a penny to haul away a truckload of the mess. "I thought it was strange he was investing so much time in a store we'd only rented for three months, but he had this passion to make things great," said Steve Komito, an early employee of The North Face. "His attitude was 'Of course it's going to work! Full speed ahead and damn the torpedoes.'"

Tompkins had no money for marketing, so he helped plan a stunt. What if an employee of The North Face rappelled down the face of a San Francisco skyscraper and grabbed a cup of coffee from the bemused building owner six floors up? Dwelle tied a 600-foot rope to the penthouse atop San Francisco's Pacific National Bank building and then, with Doug's climbing friend Galen Rowell, Dwelle slid down a rope to the sixth floor, stopped, and drank the coffee. They had alerted photographers—the media went nuts. Their rebel advertising worked as the San Francisco *Chronicle* ran a feature story. To further promote his new business, Tompkins placed an ad in the Yellow Pages and ordered 5,000 boxes of matches, each one printed with the legend "The North Face, 308 Columbus Av."

"He ordered 5,000, thinking he was ordering 5,000 matchbooks, but 5,000 boxes of matchbooks showed up," explained Dwelle. "He had all this investment and the Swiss [Ski Shop] guys tell him, 'We're going to move downtown; you can come with us if you want.' Doug replied, 'I can't move; the season has just started. I've got advertising out there; people are coming in.' And they said, 'If you want to stay here you can take over the lease.' That's how The North Face became a viable store."

Edgar Boyles, a friend of Doug's, who lived nearby and studied photography and filmmaking, visited his buddy's store to marvel at the latest upgrades. It was all handmade, more sweat than equity. "It was hard to understand what he was doing," said Boyles. "Everything in the store was done to the aesthetic level of an art piece."

As The North Face entered its second year, Doug capitalized on a deep cultural shift. More and more urban and suburban residents were taking to the mountains. Middle-class Americans had a rise in leisure time, while widespread auto ownership enabled families to take camping trips into little-known national parks. Hiking and camping became booming national pastimes. Visits to US National Parks soared from 53 million in 1958 to more than 130 million in 1968. Tompkins acquired distribution rights to novel ski technologies, including the Lange plastic ski boot and ultratight stretch ski pants. Macho climbers, ski-bunnies, wandering poets, and the odd tourist all mixed it up inside The North Face, practically guaranteeing a lively scene. The shop was often more like a living room cocktail party than a retail experience. Doug loved the show. He was fascinated by circus performers and saw his small store as a theater to be decorated, and as master of ceremonies he was never bored.

Tompkins knew his products had appeal, but he was limited by the dimensions of his tiny establishment. Yvon ran a mail order catalog for Chouinard Equipment and described to Doug how he shipped box after box of his climbing hardware. With Dwelle working nonstop, Doug organized his own mail order catalog. Instead of a typical appeal to the wallet, this was an appeal to the heart. Printed on extra-large sheets of paper, The North Face catalog featured pencil drawings. There were no photos and no models, and the catalog began with a handwritten letter signed by Douglas Tompkins. The catalog was elegant and brash. It projected a rebel attitude with a counterintuitive message—"Pack less and enjoy more."

When The North Face catalog was ready, Tompkins printed 10,000 copies and mailed them out. Two days later, a truck from the US Postal

Service pulled up and dropped off fifteen mailbags. The mailman was not happy. "You've got to sort 'em," he ordered, and left. Tompkins ignored the advice. "We're not going to do that," he told Dwelle. "We'll just put 'em in piles, order the ones on top, make it look like they're all sorted."

Tompkins and Dwelle sorted the first few catalogs by zip code and put the 10,000 unsorted catalogs back in the mail. Two days later, a US Postal Service truck pulled up again and dropped off all fifteen bags with the 10,000 catalogs. The message was clear: if you don't comply, the next time they will be destroyed. "That happened on a Friday, so on the weekend we went down to [San Francisco's] Washington Square Park, and we spread out all 10,000 catalogs in the park, and hand-sorted the addresses," explained Dwelle—"the two of us, walking around, looking at zip codes, until we got them in order. During a sunny day, we took over half of the park. People would stop and ask, 'What are you doing? Can I have one of those? Is it a promotional event?' No. But it turned out that way. They finally went out in the mail, but they may have gotten more exposure through being spread out in the park on that sunny day."

Thanks to their standout catalog and Yvon's guidance, The North Face boomed. Orders flooded in. Sales of Yvon's climbing gear also soared. Dwelle regularly drove to Southern California, returning with a van sagging on its springs from the weight of hundreds of pounds of Chouinard products. As a subculture, climbers were suddenly cool and, like musicians, they attracted groupies. Rick Ridgeway, a friend of Doug's, was in the store one day when a yellow Porsche squealed to the curb in front of The North Face and a woman's voice echoed out: *"Dooooooooooooouuug!!!! Let's Gooooo!!!!!"* It was a rowdy and deep voice, and "Doug climbed in and she sped away," said Ridgeway. After several blocks, Tompkins was terrified. He always prided himself on driving fast, but this was too much. He jumped out of the Porsche when the driver slowed and walked back to the store. To his friends he vowed never to get into a car again with Janis Joplin.

Sporting a beard and curly brown hair, Tompkins pursued a broad understanding of free love as he rocked to Jefferson Airplane at the Fillmore and hung out with Jann Wenner, who urged him to finance a rock 'n' roll magazine he planned to call *Rolling Stone*. On weekends, Doug joined fireplace chats at the Berkeley home of mountain climbers Francis and Mary Farquhar with activist David Brower, who as president of the Sierra Club led campaigns to create Redwood National Park and Point Reyes National Seashore. Brower was a talented speaker with a resounding baritone and a magnetic personality who preached the need for mountaineers to share the spirit of the journey. "It is not variety that is the spice of life," Brower riffed. "Variety is the meat and potatoes. *Risk* is the spice of life. Those who climb mountains or raft rivers understand this."

Sitting in a living room and listening, Doug felt he was "at the feet of the Archdruid and the anointed circle." Tompkins felt a deep affinity for the one-liners and sloganeering that the tenacious Brower wielded. "He was good at boiling down an issue. He could quickly see strategies to get policy changed or stop something—an unwise dam, bad forest practices. He persevered and kept the pressure on. He didn't give up and he hated to throw in the towel."

Although they each had committed to forming a family, Yvon and Doug frequently escaped the responsibilities of running Chouinard Equipment and The North Face. They went hiking, climbing, and camping for weeks, even months at a time, disappearing into the wild. Following every extreme expedition, the two returned with complaints about shoddy gear. Why, they asked, was it so difficult to find high-quality equipment? Couldn't anybody make a sleeping bag that dried quickly? How about a tent that didn't catch the wind like a kite?

Chouinard noticed that the standard ice axe design was flawed. When he swung an ice axe into a block of ice, half the time it bounced off. Couldn't someone make a curved head that matched the arc of a mountaineer's swing? They dreamed of equipment that wouldn't fail them.

In October 1966, Tompkins planned to launch The North Face's new winter season. He needed a stunt to make a splash, so he asked Jerry Mander, a music promoter friend, to jazz up the evening. Why not have a party? Live music, ample marijuana, and cold beer could work. They heard about an up-and-coming band, a shaggy quartet that included a bearded guitarist named Jerry. "They call themselves 'The Grateful Dead.' You wanna try and get them?" Mander asked. Tompkins replied with a smile. "The Grateful Dead—that sounds good."

The Grateful Dead set up stage in front of an eight-foot-high Ansel Adams landscape photograph and played a set. Susie greeted everyone, and posed for a picture with Ron "Pigpen" McKernan, the blues singer leading the band. Joan Baez showed up. Her sister Mimi Fariña sang. The crowd overflowed into the street. To keep an eye on the ruckus, Tompkins hired motorcycle gang the Hell's Angels who sported wild beards, weaponized chains, and a reputation as outlaw enforcers. When the gig ended, Tompkins invited the Hell's Angels and The Grateful Dead to Vanessi's, a white-tablecloth restaurant. Doug and Susie couldn't stop laughing at the cultural clash between the Italian waiters with bow ties and the greasy motorcycle gangsters with sleeveless leather vests.

After the Grateful Dead concert, The North Face was firmly on the cultural map. Tompkins expanded the mail order business and opened stores in Berkeley and Palo Alto. The catalog garnered rave reviews, and The North Face store became a cultural attraction—like an art gallery stuffed with a crowd of beatniks, dirtbag climbers, and Doug and Susie's circle of friends.

By the early 1960s Kerouac's prediction of a "Rucksack Revolution" was reality, and Doug had carved out a niche in the burgeoning outdoor clothing and gear market. Catalog sales grew month after month, and Doug was besieged with requests to franchise. Yet every time he finally abandoned the city and escaped to Point Reyes National Seashore or took his daughter walking in Muir Woods or went climbing in Yosemite

National Park, he ended up besieged by friends and even strangers who needed answers to their questions about gear. They all knew Doug was atop the latest innovations and trends. "I was going nuts," said Doug. "And I thought—*What am I doing? Here I am in a sport that I love, but I end up having to talk equipment with everybody?*"

Tompkins decided to sell The North Face. He had a few debts, but, more importantly, he needed a fresh mission, an expedition, an escape. He craved ventures that broke all the rules and carved a fresh path. In 1967, Tompkins sold The North Face for a paltry $50,000. He'd launched a brilliant concept, fueled the brand with adrenaline, and was now ready to jump into something completely different. Tompkins split the cash into two ventures. He would fund the Plain Jane dress company that his wife, Susie, and her friend Jane were slowly building. With their own creative juices, a bit of cash, and some marketing tips from Doug, they could expand their fledgling clothing business. Perhaps even hire a pattern maker. The business was clearly working. Jane modeled in Paris and brought back the latest fashions. Susie and she together knocked off the designs, fitted the sizes to American women's bodies, and sold their dresses at flea markets. Doug put the second chunk of money into his dream to film his own adventures. If he could get paid to climb, hike, and explore, what could be better?

Bruce Brown had just directed *Endless Summer*, an ode-to-surf-film that cost $50,000 and went on to gross $30 million. Across the United States *Endless Summer* broke box office records for a travel documentary. In Lawrence, Kansas, amid cornfields, and as far as a surfer could be from Eden, college students and farm boys waited in line to imbibe Brown's fantasy surfscape. Perhaps, Tompkins thought, the audience for adrenaline is limitless? Could he chronicle his passion for extreme climbing with the same irreverence and humor that Brown brought to surfing? When he put The North Face up for sale, the asking price was the budget of *Endless Summer:* $50,000. Now Tompkins could reinvent himself as a movie director; all he needed was one wild idea.

In the spring of 1968, Doug fell in love with a photo, and his life compass flipped from Northern Hemisphere to Southern Hemisphere. He couldn't take his eyes off the statuesque silhouette he first spotted in a black-and-white photograph. Doug stared at the image—there was no doubt these were perfect lines. A singular beauty. He revved his black Triumph and, unable to contain his glee, roared his motorcycle south along California's picturesque Route 1. Leaving San Francisco behind, the wind whipping his hair, he slalomed like a skier. He banked through the curves—redwood forest to his left, a cliff to his right, and the Pacific Ocean crashing into the rocks below. Doug roared south to share the news with Yvon—he had a new *amore*.

Doug found Yvon at his beachside hut in Southern California hammering pitons in his workshop. They surfed and partied on Ventura Beach. Doug talked passionately about his latest attraction. He showed Yvon the picture; they both stared. Doug was obsessed. He couldn't get his mind off the image: Mount Fitz Roy, an 11,289-foot mountain peak rising above the flowing Argentine prairie grasses and sculpted like an arrowhead. Surrounding peaks were so steep they looked like knitting needles balanced on end. The granite walls were nearly vertical. Buffeted by brutal winds and buried in deep snow, Fitz Roy was extremely challenging. Only two expeditions had reached the summit. Not a single American climber had reached the Fitz Roy peak. *Alpine* magazine summed up Fitz Roy as "good ice climbing, weather unstable, wind ferocious."

The photo that so inspired Doug and Yvon was the back cover of a mountaineering magazine. Staring at the photo, Doug and Yvon understood the challenge. Spears of granite rock—similar to the climbing conditions they knew so well in Yosemite—but here at the southern extreme of the Americas, they would be closer to Antarctica than the equator, literally at the ends of the earth. The beauty of Fitz Roy was as impressive as the technical challenge. Six needle-like spires formed a landscape any climber could love.

Although barely one-third the height of Mount Everest, Fitz Roy is arguably more challenging. The final 2,000-foot climb up Fitz Roy is as steep as the Empire State Building. The bulging granite wall carried the heft of Rio de Janeiro's iconic Sugarloaf Mountain, but Fitz Roy was girdled by glaciers, hammered by gale-force winds, and dressed in ice. The mountain had long been shrouded in mystery and myth. Until 1908 cartographers assumed the whirling clouds around the peak were puffs of smoke and dutifully mapped "Volcano Fitz Roy."

Situated at the southernmost tail of the Andes Mountains in South America, Fitz Roy belonged to Patagonia, a land as mythical as it was tempting. Doug knew that Yvon was among the few climbers skilled enough to scale Fitz Roy, but could he lead a full expedition to summit? Could they film it? They both dropped all their plans and that morning began plotting an epic journey. How to climb the unclimbable? What time of year was best? Who else did they need to confront this challenge?

Yvon was *in*. Earlier that year he had read a chronicle by the French climber Lionel Terray, describing the first ascent of Fitz Roy. Terray described climbing as a noble pursuit existing outside the realm of reason and sensibility. Climbers, he said, were *Conquistadors of the Useless*. Fitz Roy, suggested Terray, was singularly difficult. The gusty winds high on the mountain made tents impractical, and days on end he lived inside hand-carved snow caves. One climber, Yvon read, lost his hearing from the constant roar of wind during the Fitz Roy ascent. Not until three months after the expedition could he once again properly hear. Another French climber died. Jacques Poincenot fell into raging waters near base camp and drowned under a rock. His teammates christened a spectacular peak in his honor and mourned that during the easiest part of the expedition—simply crossing a river—Patagonia had stolen away their mate. Terray claimed that of all the mountains he climbed, Fitz Roy was one of just two that he never again wanted to face.

Surfing and brainstorming in the heat at Ventura Beach, Yvon and Doug sketched out a six-month-long expedition. "We approach our mountain by land, driving from California, surfing on the west coast of Central and South America, ski for a month in Chile, and then on to Patagonia," Doug gushed to Yvon. "We are going to 'Hog Fun' as much as we can for six months." Tompkins baptized their adventure "The 1968 California Fun Hog Expedition to Patagonia."

June through October snowstorms battered Fitz Roy, so Doug and Yvon planned to arrive in November—before summer winds picked up. But the window between late spring blizzards and early summer gales was fleeting and dangerously unpredictable. They planned their departure for July and, given the inverted seasons in the Southern Hemisphere, they would drive from summer straight into spring. "Plans were piled on plans, fantasy on fantasy; by nightfall we had concocted the trip of trips," said Tompkins. "We were like boys who sneak into an ice-cream shop at night to make themselves a gigantic sundae or banana split; it's all free and the soda jockeys are always so stingy with the syrups."

Chapter 2

Conquistadors of the Useless

It's hard to say sometimes why I'd get in a van and drive 16,000 miles to climb some mountain. I never really thought about the motives. I never really sat down and analyzed why I was going to do that. It would probably scare me.

—DOUG TOMPKINS

Just weeks after his beachside brainstorm with Yvon, in July 1968, packing what little remained of the $50,000 cash from selling The North Face, Doug Tompkins stuffed a '65 Ford Econoline van with two Bolex 16-mm cameras, twelve rolls of film donated by mountaineer David Brower, a pile of spare tires, climbing ropes, downhill skis, and wet suits. Leaving San Francisco was easy for Doug—and painfully uncomfortable for his wife, Susie. She was nine months pregnant with their daughter Summer Tompkins and was already caring for two-year-old Quincey Tompkins. She was busy cofounding Plain Jane, a startup clothing company. As soon as baby Summer Tompkins was born, her father hit the road. "He basically gave some money to Susie and said, 'Okay, you've been talking with your friend Jane Tise for a long time about these ideas for your own clothing line. Well, here's some money. Why don't you go and do it? That will give you something exciting to do while I'm gone on this trip,'" confided one friend. "It was very much a deal to make leaving his family not really easier but possible, without a sense of abandoning them."

Doug drove south out of San Francisco with Lito, his friend and fellow skier when they lived at Squaw Valley near Lake Tahoe. Lito, a native of Bolivia, spoke Spanish and was hired as cameraman for the climbing adventure. Lito felt a deep bond with Doug—they were both passionate about life as climbers. "You literally put your life in your partner's hands—the hands that hold the rope that will save you if all goes wrong, if you fall," said Lito. "This immense trust leads to an immense bond."

In Ventura, California, they were joined by Yvon Chouinard and Richard "Dick" Dorworth, a muscled giant and a bookworm who scribbled furiously in his diaries. Dorworth was among the best downhill skiers in the world. He'd raced with Olympian Jean-Claude Killy and six years earlier had set the world record for speed skiing by topping 106 mph. A fifth climber—Chris Jones—was already in South America scaling some of the most difficult peaks in the Andes. He'd join them en route—location to be determined.

In Los Angeles, the four "Fun Hogs" chopped off their beards, tidied sideburns, and trimmed moustaches. Mexican *federales* were less likely to detain and jail four gringos in a van full of camping gear if they were clean shaven. For the same reason, they took several vows: No pot. No pills. No LSD.

Yvon designed and built custom shelving into the van to hold their climbing ropes, camp stove, skis, and winter clothing. He roped the surfboards to the roof. Their sequel to the surf movie *Endless Summer* was on schedule as they launched their monthslong trip to climb the unclimbed, surf the waves never surfed, and ski the snow-covered slopes of smoking volcanos in a quest to reach this mythical land they knew only from a few scattered photographs and campfire stories: *Patagonia!*

In the summer of 1968, the United States was embroiled in a divisive culture war. Three months before their departure, in April, James

Earl Ray assassinated Martin Luther King Jr. in Memphis. The ensuing riots left dozens dead and the national psyche scarred. In June, Sirhan Sirhan murdered Robert F. Kennedy in Los Angeles as the charismatic former attorney general celebrated his victory in the California Democratic presidential primary. As RFK prepared to inherit his brother JFK's mantle he instead joined him in death, as another great hope gunned down. Fitz Roy seemed the perfect antithesis to US madness. It was 16,000 miles south along the Pan-American Highway, a road that, in parts, was barely more than a well-trodden cattle path. Maps of the region were often in Spanish and sparsely decorated with a road, a symbol marking a gas station, and perhaps the occasional Mayan or Incan ruin to visit.

Two Fun Hogs sat in the front seats and two Fun Hogs in the back. Twenty-five hours of cassette tapes guaranteed that the Grateful Dead, Jefferson Airplane, Joan Baez, and Bob Dylan would serenade them along their monthslong southern odyssey. A Nagra tape deck with speaker was rigged up to keep the van's driver awake on long night shifts.

Crossing the frontier into Mexico, the Fun Hogs drove to San Blas, a fishing village on Mexico's Pacific coast. Lito set up a shot, shooting from a dune as he filmed Doug and Yvon surfing side by side. It was an uninspiring knockoff of *Endless Summer* and void of originality. Tompkins blamed Lito and ordered his cameraman about like an extra on a set. The Fun Hogs told Tompkins to back off. They had thousands of miles and hundreds of beaches to film. Tompkins found it hard to relax unless he was in overdrive—"as if he needs to move to prove to himself that he is still alive," Dorworth wrote in his journal.

In Mexico City, Tompkins found a photo lab to develop the first reels of film. He planned to ship the 16-mm masters back to the US. In case of any mishaps, the master footage would be safely stored in California. When Tompkins received the developed rolls, not a single frame was in focus—everything looked jittery, as if the camera was shaking. All their footage, including the opening scene leaving Los Angeles, crossing the border into Tijuana, and surfing in San Blas, was useless. Lito opened the camera and found that a pressure plate wasn't holding the 16-mm film in its guides. They reset and launched again—next stop, Central America.

Tompkins fell in love with the street food as they drove through villages, each with its open-air marketplace. Although he spoke only rudimentary Spanish, Doug wandered the markets, exploring the wealth of handwoven fabrics, local ceramics, metalwork crafts, and a culture in which handmade articles were a way of life.

In Central America, Tompkins loved the bustle, the colors, and the spicy, exotic food. Fresh fruit was so abundant it seemed free—a pineapple cost a penny. Driving day and all night pushed them ahead of schedule. Dorworth was a machine behind the wheel. He loved to drive so much at night and seemed so immune to fatigue that Yvon wondered if Dick was secretly micro dosing LSD to keep awake.

The van's weak headlights and Lito's worse eyesight made it impossibly dangerous for him to drive at night. Even during the day, Lito was a terrible driver and used only for extreme circumstances. Doug himself drove too fast, racing the engine and bouncing his buddies' heads against the van's metal roof, eliciting a rain of insults. "We gotta get there," he would mumble in a tone that sounded both unsympathetic and arrogant—as if he were ignoring them.

In Guatemala, Doug visited a fortune teller. The fortune teller's booth consisted of three wooden bird cages and in each pen a small pile of fortunes rolled up on small pieces of paper like scrolls. After paying the nominal fee, Doug watched as the street hustler whistled to the bird, a signal to fetch a random scroll. Unrolling the fortune while Lito filmed, Doug was struck by the message—*Tu Familia Te Busca* ("Your Family Looks for You").

In the hills outside of Antigua, Guatemala, in the predawn light, the Fun Hogs were ambushed. While sleeping on the ground next to the van, Tompkins awoke to the metallic clack of a bullet being chambered.

"I have my sleeping bag pulled up around my head, only my eyes in the opening," he wrote. "I opened one eye to see what was up. A teenage soldier was pointing his automatic rifle at us, shouting in Spanish. The point of his gun is shaking as he spoke."

Lito, the only Fun Hog with conversational Spanish, raised his hands over his head. Speaking rapidly with the young soldiers, Lito understood that they were looking for a man they had shot the night before but who had escaped—a wanted and wounded rebel. Each Fun Hog was ordered to slowly exit his sleeping bag to prove he was intact, with no bullet wounds. Lito explained they were adventure tourists on a vacation—pointing out their blue California license plate as evidence. The Fun Hogs assured the soldiers they would report any suspicious activities. As soon as the soldiers left, the Fun Hogs broke camp, packed their goods, and hit the road. "Those were the eyes of a killer," Dorworth wrote in his journal. "I have no doubt he wanted to shoot us."

The men drove nearly nonstop to Panama, where the road abruptly ended. From Alaska to Tierra del Fuego, the Pan-American Highway allows drivers a single route—except this stretch, known as the Darien Gap. Tropical diseases, a swamp, a daily deluge of rain, political differences, and a general lack of land-based traffic made this tiny neck of southern Panama a no-man's-land. How would they navigate the eighty-mile gap? They reserved a ride to Buenaventura, Colombia, but the ship never arrived so they bought passage on a Spanish freighter headed to Cartagena, on Colombia's Caribbean coast. The change of itinerary added 1,000 miles to their trip, and Colombia was in the early stages of civil war.

Fearing ambush from guerrillas or government forces, the Fun Hogs beelined to Ecuador, where Yvon's briefcase was promptly stolen. The thief darted away with their paperwork, and, more damaging still, eight rolls of film they'd shot. Arriving in Peru, Tompkins was worried. Their trip was on schedule, but usable film footage was scant. Their film was in trouble. In Peru, thieves smashed the van window and attempted to steal climbing equipment and the little cash remaining. The Fun Hogs sold their surfboards for gas money. They also picked up Chris Jones, an accomplished English rock climber fresh from summiting peaks in the Peruvian Andes.

As they left Peru and crossed into Chile, Doug and Dick felt a wave of relief. They knew Chile from their ski training trips and found the thin, spaghetti-shaped nation to be a well-run, democratic republic where "things worked" and violence was rarely a worry.

Northern Chile is defined by the Atacama Desert—among the driest ecosystems on Earth. In areas of the desert no rainfall was recorded for over a century and daytime temperatures soar above 100 degrees Fahrenheit. At night the Atacama turns chilly and, with almost no moisture in the air, holds bragging rights to the clearest skies in the world and a dazzling nightscape of stars.

Four centuries earlier, during the Inca reign, runners sent by the rulers in Lima ran across the entire desert. Oasis to oasis, the runners carried messages to the southern reaches of the Incan empire in central Chile. The Fun Hogs debated whether their Econoline, increasingly prone to mechanical breakdown, could match the Incas and survive the 700-mile journey across the uninhabited desert. Of the van's six cylinders, only three still fired correctly. And the Bob Dylan sticker on the back bumper became so faded that they could barely read the message—"Don't Look Back." Yvon oiled, babied, and tinkered with the engine, but it was clearly dying.

Near the northern Chilean city of Iquique, the Fun Hogs camped on a cliff high above the Pacific Ocean, hoping to spot whales migrating near the coast. Tompkins was asleep when he felt the van lurch forward. "With neither warning, nor provocation, the van slipped out of gear and rolled toward the boys sleeping on the ground, and to the cliff and certain death," he later wrote. "I managed to dive over the seat and plunged to the front floorboards, and put on the brakes with my hands, just managing to stop the van before it ran over the fellows and the cliff beyond."

Two days later the Econoline limped into Santiago. Yvon unscrewed, unbolted, and dissected the engine. He visited lathe shops until he found one willing to bore out the Ford engine block; then, in a weeklong show of engineering prowess, he cleaned and rebuilt the eighty-five-horsepower engine. While he and Doug labored, Chris Jones, Lito, and Dick Dorworth explored the local flea markets, purchased ropes and climbing supplies, sampled local ice cream, and observed the street marches.

Driving south out of Santiago, their pace slowed to twenty miles per hour as the dusty Pan-American "Highway" turned into a slippery, muddy rut the color of the six-foot-thick topsoil that locals praised as "chocolate." Changing tires was easy compared to pushing the van out of the muck. They camped surrounded by thick forests as they read aloud *On the Road* by Jack Kerouac and *Red-Dirt Marijuana and Other Tastes*, a collection of short stories by Terry Southern.

"The Chilean countryside is amazing," Dorworth wrote in his diary. "It is reminiscent of Oregon and Washington and British Columbia but suddenly it is unlike anything I have seen before. Primeval. Lonely. A harmony in patterns and colors of green. Something left over from a more innocent, therefore wild time of the earth."

Doug stared at the landscape, which over the course of a single day's drive had morphed from lush green paradise to gray wreckage. Settlers known as *colonos* had hacked away thick forests, clearing land for cattle and homesteads. Roadside stumps and trees sawed off at shoulder height were aged remnants of a massive forest fire. Thousands of trees lay toppled on the ground; it appeared as if a brutal wind had slapped down the entire forest. Pilots flying over the area said it looked as if a giant had scattered thousands of matchsticks across the countryside. Some farmers gathered the scattered trunks and used the wreckage as firewood, but in many areas fields of jagged silhouettes stood tall, gray ghosts from what Dorworth surmised was "some great fire of another age."

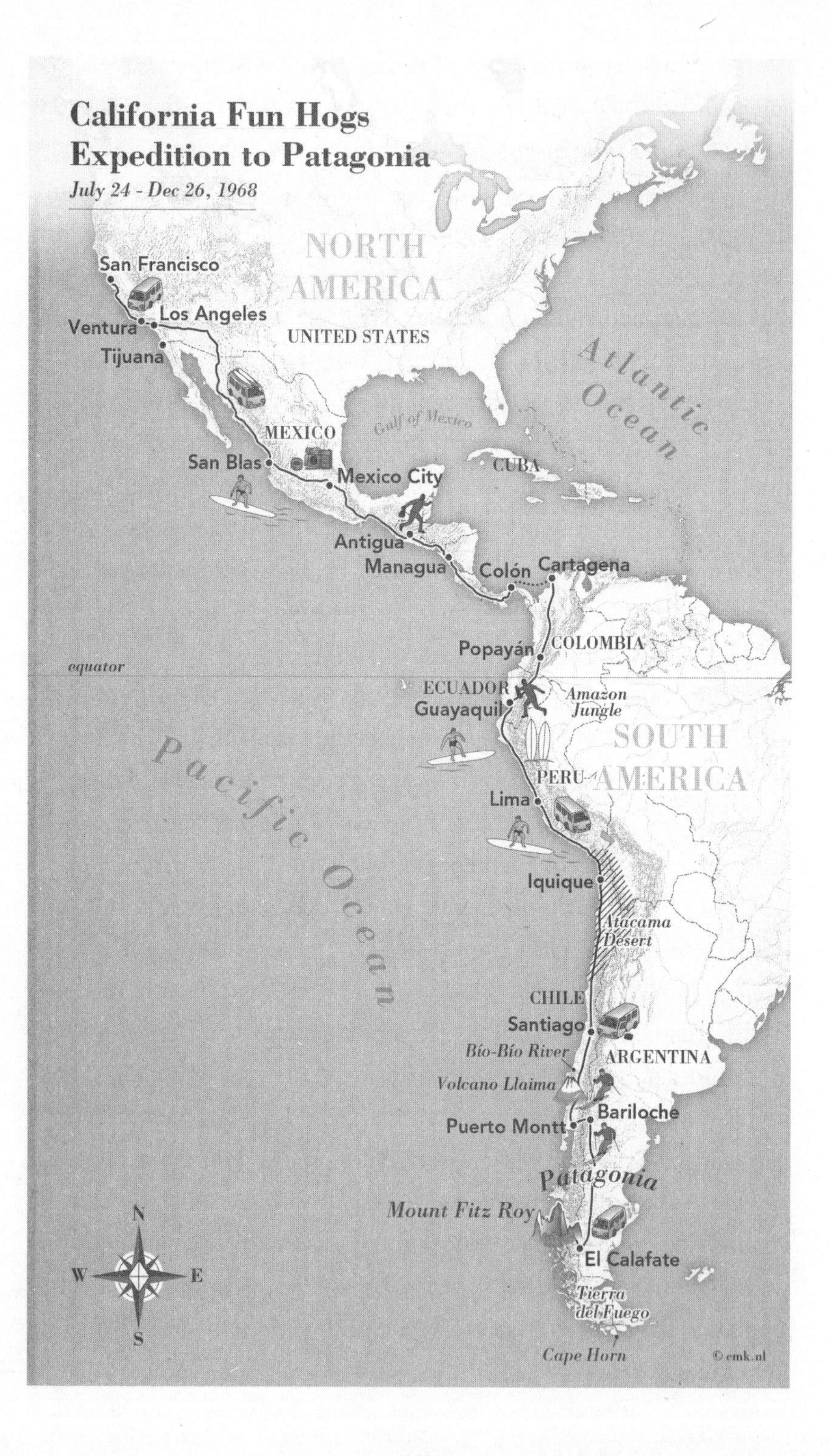
California Fun Hogs
Expedition to Patagonia
July 24 - Dec 26, 1968
NORTH AMERICA
UNITED STATES
San Francisco
Los Angeles
Ventura
Tijuana
Atlantic Ocean
Gulf of Mexico
MEXICO
CUBA
San Blas
Mexico City
Antigua
Managua
Colón
Cartagena
Popayán
COLOMBIA
equator
ECUADOR
Guayaquil
Amazon Jungle
SOUTH AMERICA
Pacific Ocean
PERU
Lima
Iquique
Atacama Desert
CHILE
Santiago
ARGENTINA
Bío-Bío River
Volcano Llaima
Bariloche
Puerto Montt
Patagonia
Mount Fitz Roy
El Calafate
Tierra del Fuego
Cape Horn
N
W
E
S
© emk.nl

Hundreds of miles south of Santiago, the Fun Hogs approached Volcano Llaima and made plans to climb to its smoking crater at 10,250 feet and then ski thousands of feet down the flanks. Covered in thick snow, Llaima looked like an ideal slope. But since poisonous fumes and smoke billowed from the crater, it was doubtful that they could reach the peak. Carrying skis, oversized ski backpacks, and walking in lace-up leather ski boots, the men trekked toward the crater—and noted the rocks were warmer the higher they trudged.

After climbing for eight hours, the men circled the lip of the crater and approached from the south. A stiff breeze blew the deadly fumes due north. "We were able to stick our ski poles out over the lip," said Tompkins. As Lito filmed, Doug and Dick skied the fresh powder side by side, carving neat turns as Yvon cartwheeled behind them—a mix of crossed ski tips, upended poles, and curses. An intermediate skier, Chris Jones kept discreetly out of the camera frame. For a week they climbed, skied, and filmed. Then, short of cash, they sold their skis for $95—a needed infusion for goods and gas.

Arriving in the southern port city of Puerto Montt, Doug went to work as a forger. He cut a deal with the owners of a Chilean print shop, gleefully explaining to the law-abiding and conservative locals that he'd lost his original paperwork for Argentine customs. He needed help creating a replacement customs stamp. Would they help? Together they could dupe Chile's rivals on the other side of the Andes: the flamboyant and pompous Argentines. Tompkins returned time and time again to improve the counterfeit stamp. The final product was near-perfect and only cost three dollars.

From Puerto Montt, the men took a tiny ferry (capacity: two cars) across a frigid lake set in a fjord framed by snow-topped mountains. It looked like a postcard from Norway. The diesel boat was barely seaworthy, and the engine stalled so often the Fun Hogs wondered if they would be marooned on the remote lakeshore. The untouched green forests appeared void of human presence. No roads, no towns, no signs of settlement. The owners, often wealthy families in Santiago who had inherited land grants, were rarely present or aware of what they owned. The soil was too damp for farming or grazing and the lakes too cold for bathing. Fishing, however, was outstanding. Eastern brook trout—the national fish of Nova Scotia and New Hampshire—imported by anglers decades earlier, had adapted all too well in the nutrient-rich waters. The aggressive trout (part of the salmon family) ate out the local fish, colonized the rivers and lakes, and as measured in biomass soon accounted for 80 percent of all regional fish.

Preparing to cross the border into Argentina, the Fun Hogs toasted their fake rubber stamp in hopes it would work. Then the men celebrated—their falsified paperwork had fooled the authorities. They had smuggled the van into Argentina without having to post a $6,000 bond, which they would have to forfeit if they sold it (which they planned to).

In Argentina they found empty roads and clear weather, so they tore across the pampa. The flat, overgrazed grasslands reminded them of the American Southwest. The land was patrolled by lone cowboys riding with bedroll and gun, trailed by a pack of dogs. Here the land was dry, hot, and empty. Much of the rain fell on the Chilean side of the mountains. Occasional flocks of wild, ostrich-like rheas trotted across the flatlands, but there were few other signs of wildlife. Desolate frontier outposts offered dozens of veterinary medicines for sheep and cattle but little more than basic provisions for humans. Chouinard spotted a shovel used to feed coal into the iron stoves so common in the region and recognized the value of the short-handled, sharp-bladed spade. Remembering the French account of building ice caves, he purchased two for their upcoming mountain climb.

Arriving in the touristy Bariloche—known to some as "the region's Jackson Hole"—they found a neat and orderly city built on the southern shores of the clear blue lake Nahuel Huapí. Bariloche felt distinctly German or Austrian, as bakeries offered *kuchen* and *strudel*. Fancy homes were built from blocks of cut stone adorned with thick wooden beams and topped in steeply slanted roofs to shed the occasional deep snowfall. "Steaks are too cheap to afford anything else and the candy is outrageously good," Dorworth noted. The men sheltered in Refugio Frei, a hut at the base of Cerro Catedral, and started climbing. "We adopted the same strict steak diet that boxers do, anticipating lean times ahead," noted Tompkins.

The climbers were terribly out of shape. Bouncing in the van, eating ice cream, and working on the engine had done little to prepare them for the upcoming challenges of Fitz Roy. Dorworth in particular needed to tune up—his entire rock-climbing experience was just a few months in Yosemite, and there was still some doubt whether he'd be invited all the way to summit. Given his superhuman strength and gloriously elegant skiing, Dorworth wondered if he was just a Sherpa and a poster boy for the movie. Training with Chouinard in the mountains around Bariloche, Dorworth learned quickly. He was a phenomenal athlete and drawn to the challenges. "Climbing teaches a man to protect himself against his own possible mistakes," he wrote in his journal.

Climbing has many, many elements to it that other sports don't offer. The risk factor causes us to judge and carefully weigh the risk against the circumstances that you're in. That is, the conditions and what that risk is worth. And to evaluate your skills and ability to pass through the risky area safely. And learn how to avoid and minimize the risks so they don't kill you. But knowing that shit happens in the elements and misjudgment can be very costly. . . . You don't know what life is until you have seen death. So you get right up to the edge. It teaches you about pain and privation and how to endure those things—there are times for intense concentration and then times for intense reflection. . . . It fit into my sense of what balanced living is about. . . . It's outside the material world of desire.

—DOUG TOMPKINS

Lito found climbing to be hypnotic. "Climbing requires one to be totally alert, totally present, totally responsible," he said. "We were lucky. Doug and I never lived through any terrible dramatic moments, never survived any near-death experiences, on our climbs. We paid attention; we were bold but careful."

Resting in Bariloche, the men bathed, shaved, and restocked food supplies. Calling home was an intricate affair involving various delays, connections by operators in Argentina and the United States. For every minute he spoke to Susie, Doug paid a small fortune, so he tried to keep the calls brief. But Susie was ecstatic as she explained that her little dress company had struck it rich. A dress that partner Jane designed was featured in a full-page ad in the *San Francisco Chronicle*. Follow-up orders were swamping the company. "I need your help!" she pleaded. Tompkins bought an airplane ticket from Bariloche to San Francisco. He packed a change of clothes into a small bag and told the Fun Hogs he was leaving.

The other men were stunned. On the eve of the key scenes for their movie, the director was leaving? To see his wife on the other side of the world? Tompkins was adamant and firm. It was important, and it would only take a week.

Tompkins faced an existential crisis as he boarded the flight home. He had no income, little savings, and two small children. For Susie, the rigors of living alone for months with their two young daughters while running a small business meant she was working triple time. Parenting was never Doug's forte. He was absent from his daughters' lives for months on end, and "frenetic" was a polite way of describing his idea of a family getaway. Susie needed help. And to complicate the picture further, Doug had promised Susie he was making an iconic film. He was nowhere near that, and knew it.

When he arrived in San Francisco, Susie told Doug the full story. Joseph Magnin, San Francisco's luxury department store, had featured a dress from its "Plain Jane" collection in a newspaper ad. When readers saw the advertisement, everyone wanted the dress. Demand soared. Joseph Magnin asked for hundreds of dresses. The $15,000 order (roughly $100,000 in 2021 dollars) hurtled Plain Jane from selling out of the back of the station wagon into a small company with tiny offices and start up staff.

Using contacts gained during his three years building The North Face, Doug launched into a crash course on clothing production. Jane Tise and Susie were working nonstop from a small, two-bedroom flat. Doug joined them and, packed together, they brainstormed. Who could sew so many dresses so fast? Doug investigated Chinatown. Susie fancied she had an idea about where to source the wholesale fabrics. Jane designed the clothes. "It was real mom-and-pop," said April Stark, who worked as Jane's assistant. "Jane would sketch a dress, and they had a pattern maker, very astute guy. He'd make the patterns up."

The kitchen table at Tise's apartment was used for cutting. In a small office above a massage parlor in San Francisco, Tompkins helped amp up the dress production. Doug careened between Chinatown, where he peered into alleyway sewing shops, and the city's financial district, where he parlayed the purchase order into a quick loan.

After a week, it was time for Doug to return to South America. The Fun Hogs were expecting him. But as the Fun Hogs packed up their bags and prepared to hit the road, Tompkins fired off a telegram. ONE WEEK MORE. STILL WORKING. ALL GOOD. DOUG. The Fun Hogs were furious. "Jones is pissed. Yvon is down and disgusted. I am disappointed," Dorworth wrote in his journal while downing beers at the Munich Bar in Bariloche. "Tompkins at times shows a profound lack of consideration for others. In this case, his friends."

When Tompkins returned to Bariloche, he hardly bothered to make an excuse. Instead he acted like he was the one in a rush. The Fun Hogs abandoned Bariloche and drove south through El Bolson, past Los Alerces National Park into an ever-more-desolate landscape. The few human settlements were pioneer cowboy outposts and massive sheep farms, with hardly more than a single trading post for miles. Slowly they drove along the rutted and potholed road toward Fitz Roy. After eleven flat tires and 180 back-jarring miles, they spotted Fitz Roy. "Our hearts rose and sunk," Tompkins wrote.

> *We weren't prepared for it. Thousands of miles of driving, three and a half months on the march wasn't enough! Had we somehow made a mistake? We hadn't known it would be like this! So big, so beautiful! So scary! This was a gigantic Chamonix, a gargantuan Bugaboos. Huge was our only impression and we were sixty miles across the plain. To the south of the range, at the end of Lago Viedma, spilling into this lake, more than sixty miles long, was a Himalayan-sized glacier! Those first few minutes were perhaps the most mentally debilitating of the whole trip. A strong fear, a sense of losing confidence, came over me, like the way you feel when coming to Yosemite to climb a big route on El Cap; you drive in and suddenly see the wall.* I'm going up there? *you ask.* Wait a minute. Should I? Can I?
>
> —DOUG TOMPKINS

Chapter 3

The Snow Cave

You never know how an adventure will influence the rest of your life. I spent a total of thirty-one days confined to a snow cave. I had skewered my knee with an ice axe cutting ice for the stove. I stayed on my back staring at a gloomy ceiling of ice inches above my face. Every time we started the stove the walls dripped onto our down sleeping bags, which became useless wet lumps. I turned thirty years old inside that cave; it was a low point in my life.

—YVON CHOUINARD

The dirt track heading around Lago Viedma toward the distant spires of the Fitz Roy range petered out at a river, the Rio de las Vueltas, before reaching the mountains. A shaky suspension bridge allowed the Fun Hogs to cross the river, but this was the end of the road for their faithful van. From here they would walk. "Pack horses are the only way to get to the mountain, and then our own two legs," Tompkins wrote. "So, we thought very carefully about what we would really need."

A chance encounter with Lt. Silveira of the Argentine Army scored them free packhorses for the trek through a national park, all the way to the forest at the base of Fitz Roy. Using the military horses they saved dozens of trips ferrying gear and provisions on their backs. Tompkins was delighted. Sporting a flat-brimmed Chilean *huaso* hat, he threw a duffel bag over his shoulders and, as Lito filmed, he skipped over the rickety suspension bridge spanning a river that already carried a tragic history. The men knew that it was here that one of the climbers in the French expedition had been swept to his death in the strong current. A sharp spire of granite beside Fitz Roy was named in honor of the drowned French climber. "Poincenot Needle" was a reminder that despite its beauty, Patagonia could also be deadly. But today the mood was light. Halfway across the bridge Tompkins jumped, grabbed a guy wire, and knocked out a set of pullups. He flashed a grin ear to ear.

One full day's hike above the river, the Fun Hogs began organizing their base camp on the same spot used by Lionel Terray and the pioneering 1954 French expedition. Tompkins marveled at the nuggets of history they unearthed, rusted cans filled with dirt, and remnants of an oven that Yvon would put to good use, baking bread. Base camp was deep in a native beech forest, just below a world of snow, ice, and rock.

The men left base camp and began to scout their route up Fitz Roy by hiking past an icy lake, Lago de los Tres, at the base of Fitz Roy's skirt of glaciers and snowfields. They found easy going up the long, steep snow slopes until they hit a barrier of cliffs, blocking access to the Piedras Blancas glacier. This was the "white highway" that would lead the climbers still higher to the beginning of the serious technical climbing. They found a notch in this first cliff band, the so-called Passo Superiore. Here the upper mountain was suddenly revealed, and there they decided to situate their first camp. On a large mountain, climbers often pitch tents to create a series of camps leading them to the top—but not on Fitz Roy. Patagonian winds would shred the strongest tents. The few earlier expeditions to the area had discovered a solution: seek refuge in caves carved into the snow banks. It took strenuous days of digging and numerous round trips from base camp, hauling loads up to the pass which provided access to the route above—but eventually the Fun Hogs excavated a cave to withstand the harshest weather.

Their next step was to cross the wide Piedras Blancas glacier at the foot of the final granite tower of Fitz Roy. The glacier required little more than a patient but careful slog. They only crossed the glacier roped up, in case one of the climbers broke through a hidden snow bridge over a crevasse. Four separate times Tompkins slipped into a crevasse, and twice he fell in over his head and was held by the rope. Fitz Roy was well defended.

The Piedras Blancas glacier ended under a massive stone buttress that jutted from the soaring granite lines of the final peak. On top of this ridge-like buttress, known as Silla, or "Chair," the climbers carved a second ice cave, a secure high camp from which they could launch their final climb. Yvon led the crucial pitch, and soon the whole team was standing on top of the Silla, gazing up at a final 3,000 feet of vertical granite, the true challenge of Fitz Roy.

Fearing a change in weather, the Fun Hogs started excavating a second ice cave. In Patagonian climbing, everything depends on the weather, which on Fitz Roy ranged between brutal and slightly less brutal. The men worked for hours to dig, enlarge, and reinforce the second cave. Without shelter, no one could imagine surviving even a single night on the exposed ridge of Fitz Roy. Pacific storms regularly piled in from the coast, then slammed into the vertical mountainside. Dorworth joked that it was the only place he'd seen snow rising from the valley and flying up a mountain! The threatening weather made the final summit push a very calculated gamble. They couldn't race for the summit until they had stocked their two ice-cave camps with days of food, cooking gas, and enough canned food to survive if a major storm trapped them up on the mountain. A week's worth of supplies in each ice cave seemed about right.

The climb was now an exercise in endurance. Back and forth they cached their supplies. Hauling loads up to the first cave through deep snow, and then on to their high base on the Silla, all the while, they filmed any promising scene. Many days felt more like a frustrating slog than an alpine adventure. Patience frayed, tempers grew short. Doug, the most highly motivated member of the group, was also the most frustrated.

Then the team unity created by sharing meals, inside jokes, the camaraderie of a cramped van, all exploded. During a filming sequence on the glacier, an argument ensued when Lito did not respond fast enough to an order from Doug, who screamed at him. Lito put down the camera and refused to keep filming. Tompkins shoved him, grabbed the camera, and marched down the mountain. Before he left, Tompkins picked up Jones's travel diary and scribbled out a goodbye letter. Their movie was over.

Tompkins was irate as he accused the team of being lazy. "Today is about as low as I ever want to be . . ." began his four-page diatribe. "I can tell you I am sick inside, really sick, not only for you guys but for me too. I lose the most actually." Tompkins mourned for more fluid teamwork. "It is just bitching when everyone is functioning and no grumbles or complaints, expeditions especially require that when things have to be done, they get done." He strode off across the dangerous crevasse field with no spare food, no rope, and taking with him the movie camera. Before sunset he was back. Chagrined, he called a group meeting and asked for opinions. Dorworth laid it out. "Revolting! The way you treat Lito—that just can't happen."

Tompkins paused and said, "Got it."

"And from that moment I never saw it," said Dorworth referring to the "Therapy Session" that the Fun Hogs held outside their ice cave. "He never treated Lito bad again. It was like he incorporated the message in two seconds. It was the most amazing transformation I have ever seen." With the team together again, and after days climbing up and down to stock their camps, the Fun Hogs needed a break in the weather—one day to make a dash for the summit.

Inside the second snow cave the Fun Hogs waited. And waited. But the weather wasn't cooperating. Every day a new storm whirled in, fed by moisture from the Pacific Ocean and the Southern Patagonia Ice Field. The men wrote in their journals, and described favorite dishes in luscious bites, even developing an all-star compilation of plates from favorite restaurants. Yvon read Joseph Campbell's *The Hero with a Thousand Faces* as Doug, Lito, and Chris played a local card game known as Truco.

Yvon, always the designer, mended equipment with his Speedy Stitcher Awl. He patched worn boots and frayed jackets and was constantly tinkering with the length of his gaiters and the sharpness of the crampons attached to his boots. Chouinard took over the cooking duties for the whole team, melting snow for water and preparing their minimally adequate meals and occasional hot drinks. Gathering water became a key task. Taking turns, they collected chunks of snow and ice. Yvon went out to chop some ice to melt for water, and had a glancing blow. "I stuck the pick of my ice axe into my kneecap," he said. "Down deep. It probably cut some tendons in there. It was really, really painful. . . . Nobody could have done anything for me . . . unless they were a surgeon."

The temperature was stuck at freezing, which staunched blood flow but did little to promote the formation of scar tissue. Tompkins and Chouinard had studied self-hypnotization and they liked to use it as a party trick or around the campfire. Now Tompkins desperately tried to move Chouinard out of the pain zone. They talked about how they'd run their lives—free, wild, and independent. As their entombment stretched out day after day, snowfall buried the ice cave entrance. "It's a half-light situation, and it's cold. It's exactly thirty-two degrees. It's wet," said Chouinard.

The biggest challenge was to wait. To kill time they told stories. Conversations were long, rambling, and covered everything the Fun Hogs could think of, but often returned to life post-trip. What lay next? "One of Chouinard's drives was, 'I got to make some money. Must make some bread. I'm just spinning my wheels,'" remembered Jones. "And Doug was giving him advice. Doug was pushing him."

Tompkins convinced Chouinard that hard goods (like ice hammers and pitons) were such great products that a single item might last ten years. Doug explained that selling shirts and pants was a repeat business. "That's why Yvon got into the 'soft goods' business, because Doug convinced him," said Jones.

Trapped in the ice cave, the Fun Hogs began to look like five cavemen. Ragged beards trailed off their jaws in wild curls and spikes. They had no showers, no soap, no deodorant, and no story untold. Pledges were made and vows taken. Among the promises made in that cave was an understanding between Doug and Yvon. They were both hardscrabble entrepreneurs who detested authority. "Whatever the business," they swore in effect, "keep full control." "Never go public. Never dilute." Business wasn't about maximizing wealth; business was more closely linked to mental health and a desire to sleep well, explore the wilds of the world, and spontaneously surf for an afternoon or a month. "Never take a job that doesn't give you at least four months off a year," Chouinard wrote. Tompkins agreed. The game was to maintain control over the business, not let it take control over you.

Chouinard thought Patagonia itself would be a great business name because nobody knew what Patagonia was in those days. "It was like Timbuktu," he said. "How many people know where Timbuktu is? In Mexico, a parent would ask their kid, 'Hey, where have you been?' And the kid, just to blow off his parents, would say, 'I've been in Patagonia.'"

No matter how many stories they told or plans they forged, the storms outside screamed on day after day. "There wasn't a single day we could climb," said Lito. "It was howling. We couldn't go up. We thought, well at least we could probably go down."

Eleven tediously long days stretched past. "We passed time patching up the cave and dreaming—of home, of food, of anything," said Jones. "Sometimes it was so bad we couldn't go outside all day. We had heard all our stories over and over; new ideas were nonexistent; life was held in suspension."

Doug studied the shrinking food stash—tins of canned oatmeal, a pot of watery bean soup, and a stack of bacon bars. They all agreed to reduce their meals, to a few spoonfuls of food. "We were eating so little food; it's incredible how little our ration size was. I think I was getting one thousand calories a day. A silly amount," said Jones. "We didn't have enough food to keep ourselves warm."

Whenever the storm halted briefly, a couple of climbers would tunnel out of their snowed-in cave to see if the weather was breaking. But no luck. Fifty-mile-an-hour gusts made it impossible to consider going up, impossible to resume their climb. But they couldn't stay in the ice cave indefinitely. After weeks of storm, their supplies, food, and gas for the stove were all running low. "Our objective was always to go down and get more. What was preventing us from going down was our fear of the weather," explained Jones. "We knew we had to go down, because we didn't have enough food, but we didn't want to be killed."

Finally, the men made a break for it. "Ninety percent of the time, we retreat from storms we could have kept on climbing through. Most times storms are uncomfortable, they won't kill you," Doug wrote. "But on Fitz Roy these storms are a different story. We had to get out."

Rappelling down the steep, snow-plastered face of the Silla, then trudging across the Piedras Blancas glacier in a whiteout, they retreated, first to the lower ice cave and then all the way back to their base camp in the forest. It was a long and careful descent. "We hate rappelling," said Chouinard. "A lot of people love rappelling. They think that's the big part of climbing—big jumps and stuff like that. We hate that because you're totally dependent on your anchor and on the rope."

Back at the French base camp, they dug into their supplies. Yvon baked bread, they dried their clothes, and, finally moving once more, the men began regaining their strength and their spirits. Yet red clouds in the morning warned them that fresh storm clouds were headed their way. Base camp life seemed plush compared with the cramped existence in their ice cave, but their thoughts were focused on what was waiting for them up on the mountain. Fitz Roy wasn't hidden by swirling clouds. Most of the time, the final cliffs and the summit high above were visible, but screaming streamers of spindrift blowing off the peak indicated that the wind was too strong to stand up in, much less climb. They waited, impatient to climb, too stubborn to give up their quest. After twenty-five straight days of storm, the weather broke. The five Fun Hogs quickly retraced their steps to the first ice cave, spent the night, and headed toward the upper mountain ice cave.

On top of the Silla they approached the second ice cave, but where was it? A snow tornado whipped a cloud around and around the granite peaks. "I couldn't see my hands," said Lito. "I could see the wind up there knocking spindrift, ice crystals, off the peaks and blowing it way out into the air. It was howling. We knew we couldn't climb."

The men searched methodically along the ridge. Snow fell but never seemed to hit the ground. The wind had changed the mountain landscape. Familiar formations were buried, and all recognizable features erased. After an hour of reconnoitering, the team still couldn't find their shelter. "We neglected to mark the entrance," said Tompkins. "The ice cave was buried under a thick layer of windblown snow."

Then, with a whoop of excitement, Chris Jones signaled he had found the snow cave. Jones, alternating with Tompkins, hacked away the ice using the steel blades of the small shovels they had purchased in Bariloche. Slowly, they chopped their way down to their underground cave. Melting snow had turned the entrance into an icy tube as slippery as a bobsled run and as hard as a hockey rink. Tompkins and Jones crawled Marine style, elbow by elbow. Tompkins pushed away fresh snow, squeezed his hand inside the cave and made a hole big enough for his head. He slid inside. One by one the frozen Fun Hogs then entered their snow home. Their finished cave was crooked and leaky—so they dubbed it "Cado Cave" in homage to Peter "Cado" Avenali, a close friend of Doug's who was so messy and laid back that anything so upended could be classified as "that's Cado style."

Their food supplies were rock hard, the floor was wet, and the temperature was stuck at a soggy 32 degrees Fahrenheit. Given the long days of the southern summer, they reckoned they would get back before nightfall—and if not, they'd sit out the night in the cave. "We were climbing until the weather stopped us," said Chouinard. "We had no forecasts, no GPS, no way of knowing what was going on, so we were at the mercy of the weather."

The wind ate away the roof of their shelter. A crack in the floor sucked snow into the crevasse that snaked into the innards of the mountain. Every time they filled the gap, it returned. It was a painstaking job to coax their frozen stove back to life, and the men spoke little. Exhausted from the trek, scoured by the windstorm, they collapsed asleep. Despite the terrible weather, the peak felt imminent. If the clouds blew through and the skies cleared, they could soon commence their summit attack.

Two days later, on the evening of December 18, 1968, Tompkins and Chouinard looked out of their ice cave and stared in excitement. The sky was clear, the stars bright. The team quickly agreed to launch their summit attempt. They packed extra ropes for the descent, and headlamps. If forced to spend the night on the mountain, they would have only the clothes on their backs. They set an alarm clock for 3:00 a.m. with plans to begin climbing at 4:00 a.m. While the others prepared their ropes and gear, Tompkins turned the alarm back to 1:00 a.m., two hours earlier. He figured they could climb in the dark for the first hours.

Leaving the cave they found the weather bitterly cold. In the predawn dark, the team reached the steep granite wall of the southwest buttress. The rock climb they had been waiting for was *on*. It would demand precise scouting on a hitherto unvisited rock face, as well as innovative climbing strategies. Having five climbers provided extra security if anything went wrong, but with only three climbing ropes, how could they move together? How could they belay each other, and follow

The Fun Hogs Summit Mount Fitz Roy

Oct. 29 - Dec. 20, 1968

Mount Fitz Roy
11,289 feet

Cerro Torre
10,262 feet

Poincenot Needle
9,849 feet

Cerro Mermoz
8,963 feet

Upper ice cave

Lower ice cave

Glacier Field

Piedras Blancas

Lago Sucia

Glacier Lake

from Base Camp

each pitch? A pattern was quickly established: Chouinard and Tompkins took turns leading. After one rope length, the leader would fix a rope for the others to follow. Jones backed up the leaders, cleaning the hardware that the leader had placed to protect the pitch. Lito was on his own, filming with a camera that was starting to freeze, and after such a long wait for a summit day he didn't dare ask any of the others to wait, or pause while he changed the 100-foot film rolls for his Bolex.

The climbers all used the rope that the leader had fixed to follow each pitch with a sliding-gripping clip called a jumar. And here Dick Dorworth, the least experienced climber on the team, had an important role to play. As the last man up each pitch, he would coil up the rope he had just ascended and pass it on to the leaders to use again. While the five men moved up the cliff, each absorbed by the moment, Dorworth would lose sight of the others, and couldn't hear them. He felt isolated, extremely nervous, and felt like an unwanted caboose. He had to keep cool and focused. "If I fall, I die," Dorworth told himself, then tried to avoid thinking about the possible danger. "I just kept telling myself *Don't go there. Don't go there. Don't go there.*"

As clouds shot over the ridge, past the summit, it was frigid until they climbed out of the shadows. Not till noon did they reach sunlight. Chouinard and Tompkins kept probing for the best route—a line of opportunity in the cold granite that was rough with useful cracks. Tompkins remarked how similar it was to the granite that the Fun Hogs loved to climb in Yosemite Valley—but much colder. Fingerless wool gloves allowed them to keep their hands warm but use bare fingertips to grip the rock. Tompkins called the simple, handwoven gloves his "finest piece of gear."

As they rose higher, the wind pushed misty streamers of cloud up the side of the walls. The temperature stayed frigid, but their optimism grew. After one last tough pitch they cleared a notch that they figured would bring them to the summit finale. "We thought we had it in the bag," said Jones. "And then suddenly there were these ice-coated pinnacles blocking our way to the summit. That meant at least a couple of hours more climbing. And there was rime [a dangerous thin layer of ice] covering the rock."

"Reaching the crest, we saw the problem: a scimitar-like summit ridge," said Lito. "We had to cross a series of ice-plastered towers, if we could. There was no clue to the size of the gap that separated us from the summit, and only one way to find out." The men kept climbing up, then around a corner, then repeat. "Doug was climbing like a man possessed," said Dorworth. "Taking iced grooves and rotten snow in his stride."

Several hours and four towers later, they had crossed the final serrated ridge. Tompkins rappelled down a pitch, climbed up a twenty-foot crack, and unroped. "My enthusiasm to see if there would be any more technical difficulties caused me to set off ahead while the others finished their pitches," said Tompkins. "I climbed until I could see we had it, then I walked out to the lip above the top of the Super Couloir and saw my companions appear and disappear in the clouds below."

One by one the climbers appeared—Yvon, Lito, Dick, and Chris. In appreciation of his camerawork (as well as the need for the summiting shot), Lito was sent first. Each man took slow, secure steps before standing on the small summit block. Sharp peaks that from the ground looked unscalable now lay beneath them. Chouinard summed up the group's feeling: "Well, now. We have earned our freedom for a while."

Tompkins grinned and savored the peak. Clouds roared beneath them, hiding their descent route and offering brief glimpses of Lago Viedma below. They could see the vast Patagonian ice fields and volcanoes. "We were on top of the world at that moment—masters, in our own minds, of all we surveyed," said Tompkins. "Suddenly all the agonies of the ice caves, the cold hands and feet, the discomfort and poor or little food, the uncertainty in the final weeks as to whether we could get enough good weather to do the climb—it all vanished. That surge of satisfaction all climbers feel when they have reached their goal was gushing from all five of us." A rising wind poured clouds in from all sides. The Fun Hogs were nervous about being caught out in what

Tompkins called a "Patagonian bomber." They turned their attention to the descent.

As the wind picked up, the men carefully climbed down, sliding down their doubled ropes, rappelling, then pulling the rope down after them. It was tricky with winds so strong that ropes sometimes flew up past them, risking a snagged line that could halt their descent. "You don't let your guard down at all, that's for sure," said Chouinard. "In fact, you purposely know that you have to really stay alert. A lot of the times these guided guys on Everest, if the guide gets into trouble up high, the clients don't know how they got there. When we're climbing, we're looking to see, *Okay, if we get a storm here, we could bail out on this little gully here*. You're always thinking; we're pretty aware that coming down can be more dangerous than going up."

The long summer daylight lingered but then their luck ran out as the wind whipped a rope, hanging it on a ledge. "Hands frozen, headlamps peering into the falling snow, our rope stuck; it was a bad moment," said Tompkins. They stopped rappelling around 2:00 a.m., and prepared to bivouac on a tiny ledge. Bivouacking is a lot less than camping; to a climber it means stopping and waiting for morning. They tied themselves to the wall by a webbing of fixed ropes. Dorworth slept with his boots on as he could no longer feel his toes. Tompkins felt their gear slamming into the stone as the wind slapped them about. They rested, and shivered. "It was freezing cold and I didn't sleep or anything," said Chouinard. "I just kind of sat."

At dawn they made a break for their upper ice cave. Their rappel ropes wouldn't stay down, the noise of wind "cracked like a whip at a circus lion," as slowly they descended. Said Tompkins, "We arrived back at our cramped ice cave at 11:00 a.m., worn out and cold, and after a victory flan dropped off to sleep until late afternoon." The lines of the script had been spoken correctly—they had climbed Fitz Roy by a new route. They had their film.

But at home, no one knew if they were dead or alive. The men were months overdue. Susie called Yvon's wife, Malinda Chouinard, and asked if they should launch a rescue. They knew few people who were capable of finding climbers in the wilds of South America and decided that after the holidays they would contact the American Alpine Club and put out an SOS for their lost partners.

Little did Tompkins and Chouinard realize that this adventure forged their lifelong friendship and prepared them each to overcome the challenges of the business world. They had one foot on the mountain and one in the idea of creating businesses that were respectful of nature. Yet, even in their wildest dreams, they could never have imagined that thanks to their serendipitous journey, Chouinard would become a force for corporate generosity as the leader of Patagonia clothing company and Doug among the greatest conservationists of his generation. Years later, after retelling the low points he lived while stuck in the snow cave, Yvon saw a deep value in their perseverance and suffering. "It honed me to handle adversity," he said. "So it was also the high point."

Chapter 4

Plain Jane Goes Mainstream

Doug is a real force, and if he sees something that's interesting to him or that's an opportunity of some sort, he'll grab it and make it happen.

—SUSIE TOMPKINS

On New Year's Day 1969, Doug Tompkins celebrated New Year's several times as he flew home. Along the staggered time zones he toasted in Santiago, Chile; and in Lima, Peru; then in Los Angeles; and finally in San Francisco he celebrated arriving home. Susie met Doug at the airport and at home, she reintroduced him to Summer, their six-month infant daughter. Doug didn't recognize the baby. Summer carried few firm memories of her globetrotting dad who left the day after she was born.

Back in San Francisco, Doug wrestled with what was next, while Lito spent months sorting and cataloging the film from the Fun Hog trip. The footage was used in two separate documentary films, neither of which became the hoped-for sequel to *Endless Summer*. In the rarefied world of climbing cinema, Lito's short film *Fitz Roy: First Ascent of the Southwest Buttress* won the Grand Prize at the 1969 Trento International Film Festival in Italy. It also garnered a small cult following.

Doug, meanwhile, considered opening a new business in San Francisco, perhaps a restaurant. Brainstorming for weeks about food in the ice cave had whetted his appetite and churned up a wealth of foodie ideas. The outdoor equipment market was booming, and so too were the sales of Yvon's climbing gear. But selling outdoor gear was not an option. Doug promised never again to contaminate his love of the wild with livelihood. Yvon climbed and continued to produce the finest tools for fellow dirtbags, but that was not Doug's style. Addicted to outdoor exploits and unconvinced that filming adventure films would ever be profitable for him, he sought a way to keep himself engaged in the city eight months a year. Under no circumstances was he living in urban areas more than that. From the first days of dating and through marriage with Susie, he'd always made it very clear that four months a year—sometimes six—he would disappear over the horizon: no permissions asked, no excuses needed. For those four months a year he was not husband, nor father nor CEO. He was just Doug, out in the bush with buddies, experiencing a time when he could test himself and embark on hair-raising trips that most people only read about in magazines.

Yvon agreed with his close friend. "If you can't have three or four months a year for what you really love, you are in the wrong job," Yvon declared. Convinced that athletes were the epitome of impassioned individuals and solid teammates, Yvon deliberately recruited surfers to work at Patagonia Inc. "It's easier to teach a surfer to be a businessman than a businessman to be a surfer," he laughed. "When you are a surfer and the surf's up, you grab a board. You don't schedule surfing for 4:00 p.m. next Tuesday."

Doug was absent from his family's life for months at a time. Raising a family never reached the intensity of his buddy trips or work passion, although he shared his spark of adventure with the girls. "There were always new ideas. *Let's go do this, let's go do that.* And there would be a spontaneous trip to Point Reyes National Seashore with his small daughters," said Edgar Boyles, a longtime friend. "We'd climb up some hill and camp for the night with the kids. Doug's aesthetic was always go light, don't be burdened by stuff. Sometimes it would be what they call *disaster camping*. You might have what you need, but you might not."

In a telling black-and-white family portrait from the early years of Plain Jane, Doug holds the hand of one of his daughters, but barely. She clutches the single extended finger offered by Dad, who is caught wearing an expression between aloof and unhappy.

With their new clothing concept up and running, Doug and Susie flew to Manhattan to further investigate the garment industry. During their meetings they heard about a hotshot salesman. Several contacts described Allen Schwartz as the best clothing salesman in town. Intrigued, Doug arranged a meeting. "I met them at this beautiful brownstone on East End Avenue owned by Leica Camera," said Allen. "He said to me, 'You should come and work for us.' I said, 'You don't do any business.' And he said, 'If you're as good as you think you are, I'll give you ten percent commission and twenty percent ownership.' I said, 'Twenty percent of nothing is nothing.' And he said to me, 'Well, make it good.' I said, 'I'll do $300,000 next week; that will be $30,000 in commission, so advance me $15,000. And when you ship the other half, you'll owe it to me.'"

Doug agreed on the spot, and Schwartz sold $400,000 of Plain Jane clothing during his first month working for the nascent company. "We sold to them all: Casual Corner, The Limited, Saks, Bloomingdale's, everybody," he said. "The business I built was all about New York—all these chains with thousands of stores. I sold every piece of goods."

With Allen pumping sales, Susie and Doug added Sweet Baby Jane blouses (in honor of the chart-topping crooner James Taylor's hit of the same name), then Jasmine Teas (imported T-shirts), and then Cecily Knits (knit sweaters) and skirts. Their specialty was knockoffs; they would spot a trend, tweak the design, add fresh colors, and copycat their way to millions in sales.

In 1972, just three years into their new venture, sales hit $8 million, and "Susie and Doug" were a hot couple. Overnight they became San Francisco style icons. Their brightly colored, uber-comfortable clothes were snapped up. Cash cascaded into their life. Doug and Susie paid off all their debts, rented a proper workspace, paid salaries on time, and still had piles of cash to invest. Tompkins used a wad of cash to buy a red Ferrari. "He got a ticket in Bear Valley driving 110 miles an hour," said Lito. "He tried to convince the judge that this was a really exceptional car and it wasn't a problem driving a Ferrari at that speed. The judge said, 'I've always wanted to ride in a Ferrari. Why don't you show me?' So, he took the judge out for a ride in the Ferrari. They came back and the judge still gave him a big fine. That was typical of Doug. He thought he could defy the rules and sway the world to his way of doing things."

In Southern California, Yvon was also striking it rich. Together with his wife and business partner, Malinda, they gambled that sturdy rugby shirts could double as climbing attire. Yvon spotted the shirts while climbing in Scotland, and the bright colors opened up a new flank in the outdoor clothing industry in which the dominant tones were light tan, dark tan, light gray, and regular gray. Yvon knew he was onto a trend after receiving a letter from a customer asking if Patagonia could please continue to produce sturdy outfits in a "non-ugly color."

Yvon had also launched a veritable revolution in the world of rock climbing with his proclamation of "clean climbing." Instead of disfiguring the mountainside by hammering in metal spikes, Yvon promoted aluminum forms that could be wedged in by hand. Calling the new chocks "stoppers" and "hexentrics," Yvon struck a chord with both his products and the environmentally sound philosophy it represented. Like his friend Doug, Yvon was not bound by traditional forms of building a business. He imported an eclectic line of products, including wool mittens from Austria and bivouac sacks from Scotland. For the logo of his Patagonia company he asked a friend to draw the jagged peaks of Fitz Roy and suggested they add color and a storm cloud over the summit for effect. The trip with Doug had cemented more than just their friendship. When Yvon needed to manufacture overseas, he used Doug's contacts in Hong Kong, and was soon manufacturing 3,000 shirts a month. Yvon was now in what they called "the rag business."

Hong Kong became a key part of Doug's booming clothing company. Doug and Susie manufactured T-shirts and knockoffs by the thousand in the British colony, where its bilingual workforce sported a work ethic that amazed Americans. Based on a tip from one of their employees in San Francisco (who had a brother in Hong Kong), Doug and Susie reached out to a young man named Michael Ying. Doug described him as "an unemployed guy in between jobs who happens to be the brother of a girl working for us in the United States. . . . He didn't have anything to do." Doug gave Ying initial orders, and, "working twenty hours a day," Ying became the force behind Doug and Susie's massive production capacity in Hong Kong. Doug was stunned. He considered himself a workaholic and overzealous, but Michael Ying was something else. Doug called him "the real reason that we're having success in producing and delivering product."

Doug needed an overall corporate name to house his brands, and his friend Lito came up with "Esprit de Corp"—The Spirit of the People. The word "Corps" made it sound like "corporation," and Doug also enjoyed tweaking the US Marine Corps by using its slogan to sell women's clothing.

Esprit sales rose to a million dollars a month in 1973, even as Tompkins held back. "Sell what we have," he told the salesforce. "That's all we're making." Rather than chase soaring demand, Doug chose to control growth to a level that would not sacrifice quality. He pruned the suitors, carefully selecting which department stores were worthy of the opportunity to sell his growing collection of brands. "There are 45,000 stores in the United States and we sell to 2,000 of them," he said. "We want the 2,000 that we sell to, to be better than the other 43,000—that's the competition." Doug saw the clothing business as a cutthroat field with thousands of companies fighting for market share. Fashion changed every three months, so guesswork was part of the game. If he could predict the future trends, print up the patterns, and stock the inventory, then when the trend hit money would flood in. "These companies are fast and aggressive, quick to respond. And that's the nature of the industry," said Doug. "We have to stay on footing with those guys. We have to be broken down into little battalions—like guerrilla factions—so that we can compete."

Esprit de Corp clothing now ranged from a miniskirt to flowing bell bottoms in a style that epitomized West Coast liberation. These were the kinds of clothes that sporty fifteen-year-old girls and more-particular twenty-five-year-old women would buy for themselves. Mainstream dictates were snubbed. Esprit de Corp was the offspring of San Francisco's counterculture values mixed with European fashion tastes. Female liberation was here to stay, and Esprit de Corp expressed not the angst for freedom but the small celebrations of individual choice. Requests to stock Esprit clothing were often rejected. "To sell my clothes," Doug declared, "you must present them on the right stage." He was fanatical in his dictates. The tag, the button, the hanger. No detail was too small.

As he'd done at The North Face, Doug leaned heavily on Duncan Dwelle to run his company when he decided to explore the great outdoors. Just days before a trip to Scotland with Chouinard, Tompkins called Dwelle and asked for his help. "Come over and talk to me." Duncan went over, and Doug said, "Errh, uhh, I got this clothing business, going to need someone to run it. You gotta come over and run it for me. I'm leaving next week," recalled Dwelle, who had moved on to another job. "So, I quit my job and I met Doug on a Saturday morning. He had stacks of envelopes, yellow notepads where he had scribbled notes about what had been sold, notes about what fabric had been ordered, and he left to go climb the Eiger, but ended up in Scotland. At that point he had lined up factoring by selling the receivables to a bank. We had crazy market acceptance."

Although he rarely complimented employees to their face, in a private interview Doug raved about Dwelle—"I've moved him around incredibly. He's a guy who can take a situation and grasp it. He makes himself an expert in a short period of time and then applies the same good principles of thinking." According to salesman Schwartz, Doug was the company visionary, but it was Dwelle who managed the entire infrastructure. "Duncan is kind of a genius type," said Schwartz. "He's not a fashion guy, he's a country guy you'd find living in a log cabin. He was the Wozniak."

Tompkins broke many rules of corporate business and challenged truisms. He hired women for three-quarters of the workforce, and made sure the company cafeteria was stocked with organic veggies, whole wheat breads, and healthy juices. He abhorred meetings, ruthlessly enforcing a companywide rule that no meeting could have more than three participants. "I try to leave a meeting with everybody feeling that they, too, would have made the decision unilaterally themselves."

Tompkins was obsessive with aesthetic details. Workers were permitted long hair and use of music speakers with hidden jacks. The office had 300 house plants and a full-time employee for their care. Workers quietly slid about barefoot or in socks to protect the wood floors.

Leaving an unnecessary lightbulb on or a cigarette butt on the ground was a sin. "Somebody had an olive oil can with all their pencils and scissors," recounted Susie Tompkins. "It was a great old Italian olive oil can and Doug came in and said, 'Get that off her desk.' I said, 'Well, that's her personal expression.' He said, 'It doesn't belong here.'"

Tompkins required "Friday Reports" from the seventeen people who reported to him. He asked his staff to produce "Quick reports that can't take longer than fifteen minutes to write and five minutes to read. That's about all I can handle." Every week he read for about an hour and a half, reviewing these reports. "This gives me a big overview. But the reports are as much for me as they are for themselves."

Ignoring his de facto role as CEO, he branded himself as the company's "image director" and was often seen pulling double back flips, high above his workers. At Esprit de Corp, the company trampoline was larger than the boss's office. "The trampoline was up on the roof, which was pretty spectacular," said Peter Buckley, the champion high diver who taught Tompkins how to spin on the trampoline. "You felt like you were up in the clouds. Occasionally you fell off the tramp, and someone tried to break your fall. On tricky stuff, we had a safety belt. He was a great athlete. He learned so fast. And he was fearless."

ESPRIT'S 10 BUSINESS PRINCIPLES

1. Know who we are and stick to it.
2. Create demand rather than supply demand.
3. Build a brand first; a business will follow.
4. Quality before quantity.
5. Co-prosperity among customers, suppliers, employees, and company.
6. The organization receives the by-product of each employee's energy.
7. Treat all colleagues as equals. Maintain a two-way communication system to resolve grievances.
8. Give each individual a chance to realize his or her potential.
9. Take pride in excellence and achievement.
10. Enrich the societies in which each company operates.

Near the trampoline was the Esprit de Corp carpentry workshop where Jim Sweeney turned out custom oak desks, sculpted picture frames, and made workstations for the sewing shop in Chinatown. Sweeney held a PhD in advanced math and had chosen woodworking and craftmanship as expressions of his philosophy that "What you do is who you are." Sweeney had impeccable taste. Anytime Tompkins wanted furniture custom built from wood, Sweeney could make it. As a team, the two men spent hundreds of hours brainstorming furniture design, rock climbing, and sharing LSD.

Sweeney said his boss reminded him of an aquarium fish named the "Jack Dempsey," in honor of the champion boxer. "If you take a tank of water, and you put two Dempsey fish in it, the way they challenge each other is by turning sideways and making themselves look big," said Sweeney. "The two Dempsey fish will struggle to dominate the space. Basically, what happens is a fish will have his zone of that tank. They will struggle back and forth in this psychological way. They don't bite each other. But they establish dominance over a particular zone of the tank." Sweeney continued: "And the stronger, more dominant fish will eventually take over almost all of the tank. The other fish will be in a tiny corner of the tank all by himself. Doug was a bit of a Dempsey fish, in that initially in a domain he'd enter and he'd go, 'Okay, I see who the players are here. That's fine.' But then he'd start to say, 'Well, I can be in this playground also.' And then he'd become more and more dominant and he'd claim more and more of the space."

Eager for more adrenaline, Tompkins took flying lessons in Oakland, across the Bay from San Francisco, in a versatile plane known as the "Citabria"— "a-i-r-b-a-t-i-c" spelled backward. "Instead of learning in some Cessna training vehicle with tricycle gear, he learned to fly, right from the get-go, on a tail dragger," explained Edgar Boyles, his friend and fellow pilot. "Soon after, he bought a plane. And he dove into flying like it was the next sport, as an aesthetic challenge."

With the ink still wet on his pilot's license, Doug took his family on the same route he'd traveled a few years before in a van: from California to the southernmost tip of South America.

Susie and Doug replaced the back seats of the airplane with a custom-cut pad, a playpen for Quincey, age four, and Summer, two. "It was the equivalent of a flying VW Bug," remembered Quincey. "There was a foam mattress in the back, and he would fly up and then he would quickly drop, and we would float up."

From Oakland they flew to Mexico, Costa Rica, Panama, Colombia, Ecuador, Peru, and Chile. Their final touchdown was Tierra del Fuego near the tip of South America. With his children holding tight, Tompkins touched the tiny Cessna down at the end of every day, improvising a campsite or family excursion. Following the saying "easier to ask for forgiveness than permission," he frequently landed on remote beaches. In Mexico, he misjudged the high tide and just in time mobilized Susie to help push the plane down the beach to a dry spot as he gunned the engine and took off. For twelve hours his family was stranded, and Doug could do little more than land in the nearest village, pick up food, package the morsels securely, then run low-level flights just above the beach as he air-dropped lunch and dinner to them on the beach below. "We used to land on beaches and get into trouble, and we'd land on fields too," said Susie. "We had some pretty scary things, but I think when you're a certain type of person at that age, you don't think about any of it being scary. It's just like *wow!*"

As a family they camped, sleeping on the ground or in the back of the plane. Telling stories and reading books as they flew, Tompkins sometimes let the children take the controls. This was a more playful version of their strict dad. For the daughters, flying was a magic moment, an escape from the "terrestrial Doug" who was more tyrant than parent. "He was never really very interested in other people, what they were doing, their lives, their family, their anything," said his daughter Quincey.

Back in San Francisco, Susie and Doug bought a large home on a huge lot, nearly a full city block, at the head of the famously serpentine Lombard Street. The oversized yard included a hot tub, a guest cottage, and a pool big enough that friends brought kayaks to practice "Eskimo rolls." The main residence was surrounded by towering redwoods. "It was like Muir Woods in San Francisco," said Rick Ridgeway, a long-time friend. "As you left the house, the redwoods framed the Trans-America pyramid, which had become an international visual icon of San Francisco."

The hot tub was not visible from the house, and at night Randy Hayes, a local teenager, sneaked into the hot tub for risqué adventures with his girlfriend. Years later when Hayes confessed to Tompkins, he laughed. It sounded like something he himself would have done.

The Lombard Street home was an oasis in the heart of San Francisco. The upper floor of the house featured a crow's nest room with a 360-degree panorama of San Francisco Bay and a postcard-perfect view of the Golden Gate. It was the most expensive lot in San Francisco, yet from the street only a tangle of branches was visible. There was little to tempt outsiders, and the Tompkins family rarely locked the gate. Doug lobbied to never use keys. If he got locked out, he scaled the walls, and at times tripped the house alarms. Despite multiple robberies of his car radio, he never locked his auto. "I'm going to get a car without a radio," he declared. "Screw it, so I don't have to lock my car."

At home, Doug was a mad perfectionist. In the kitchen he decorated a shelf with can after identical can of tomato sauce, each label showcased and in perfect alignment. He loved the elaborate flower arrangements placed throughout the house and regularly added to his valued art collection, including paintings and sculpture, with precise lighting like exhibits in a museum. His daughters were forced to accept his strict definition of "art." "They couldn't put things up on the walls. He decorated their rooms the way he wanted their rooms to be—all gray and white," said Susie Tompkins. "He wanted to control you."

An exhibit of Amish quilts at the Guggenheim Museum in Manhattan left Doug and Susie bedazzled. She marveled over the fine stitching and studied the hues of natural dye. Doug grew enamored with the geometric perfection and began collecting the quilts, many of which he framed as wall decorations. First at home and then throughout the Esprit office, the quilts of Lancaster County, Pennsylvania, became a hallmark of the couple's collective aesthetic. When he mounted the quilts, Doug fussed over the width of the frame and the angle of the lighting. It had to be just so, all the flaws removed. The quilts provided a stage on which he could showcase a piece of perfect art, failing to see that it is often the flaws that show humanity.

On weekends, airplane trips continued as a family pastime. For lunch they would fly to The Nut Tree in Vacaville, landing at an airstrip that was part of that popular restaurant's experience; a kiddie train ran from tarmac to dining room. Three-day escapes to Cabo San Lucas in Mexico became a regular weekend outing. They explored Northern California extensively. "San Francisco is so foggy, a lot of the time you'd take off in the fog then go up and be above it all. I remember being over the Bay Area and I could see the top of the towers of the [Golden Gate] bridge. Those red towers, the top little bit," said Quincey. "We'd go up on the Fourth of July and look down and see the fireworks and we would fly under the bridge."

With Summer, his second daughter, Tompkins was tough. He had skipped out on her first six months and never seemed to have fought to overcome that emotional deficit. "Quincey was his favorite child because she would go upstairs in her room and play quietly," said Susie. "Summer would be downstairs in the kitchen in the midst of everything, either stirring the salad dressing or standing up to Doug."

On one infamous Christmas morning, Tompkins bought his children a stupendous pile of presents. Like a guilty absentee dad, he splurged his daughters with gifts. "We unwrapped them all, and he said, 'Okay. You're going to pick one, and we're going to take the rest to an orphanage.'" Said Quincey, "I certainly tried to raise my kids with the maximum amount of sense of responsibility for what you have and for those who don't have things and to teach them to be generous and charitable and selfless. But, what my dad did? That's a really aggressive way to teach that lesson to a four-year-old."

Doug's penchant for extramarital affairs struck deep into the family. Girlfriend after girlfriend passed through their lives—visible as tension between Mom and Dad. "He was a womanizer from the very beginning. He always had girlfriends," said Susie in a frank interview. "It was very hard. And I am a resilient being. I can't tell you what I went through. You have no idea."

Doug ignored the fireworks in the family and was often aloof and self-centered, acting as if being nice was beneath him. Was his stymied run for an Olympic ski medal the cauldron of his hyperdrive? Was sleeping just a few hours a night a sign of frustration or merely a by-product of his kinetic energy? Unable or unwilling to articulate the source of his anxious behavior, Tompkins was a trailblazer and a man who didn't look in the rear-view mirror. He wasn't competing against anyone else; he was racing against himself.

Despite the marital turmoil, Esprit de Corp kept Doug and Susie so busy that it was often easy to let family issues slip by as year after year, sales boomed. Esprit needed more employees, more hands on deck. Tompkins went on a hiring spree, but in his own quirky way. Instead of selecting resumes, he selected those from a certain lifestyle. He wasn't looking to continue anyone's business plan; he was whistling hand grenades into the staid department store culture. Tompkins didn't want MBAs, he wanted free-thinking adventurers. "I want to know what your taste is," Tompkins said when asked to describe his hiring guidelines. "What you do and how you lead your life. I like to meet your parents. I like to go to your house and see what it looks like. It's very easy, because you see their bookshelves, their record collection, you see their taste in how they fixed their house. You can tell about personal style from how they're dressed, but when you go to their house it extends into a lot of other things. It's like hiring hearts and not minds."

Having no experience in retail was considered a plus. "He once told me he wouldn't hire an executive unless he can prove he's gotten several speeding tickets," said Fred Padula, a climbing buddy. "He felt that the person who had the personality to get speeding tickets was the kind of person he was looking for, who could work for him."

As Esprit de Corp boomed, Tompkins went on a climbing trip with Chouinard that nearly killed them both. Climbing without ropes and buffeted by heavy wind, they were trapped by an ice storm in a Scottish nature preserve. Tompkins was nearly blown off the mountain slope. Then an avalanche destroyed their bivouac campsite halfway up a mountainside. "We'd been sleeping there and several hours earlier had come down to a base camp when we heard a crash," said Chouinard. "Looking up we saw the entire ledge where we had camped snap off and disappear. . . . Doug and I were both overly cavalier about jumping into dangerous things. It's almost as if we've had a subconscious wish to invite catastrophe so we can try to get out of it."

Tompkins thrived in these expeditions where the goal was to travel to the edge of the map, then continue indefinitely, over the horizon. Women were not banned, yet the trips carried the stamp of a boy's club. Suffering was to be expected. Complaining not allowed. No one depended on another to live. It was survival not of the fittest but of the fit.

It was also the golden age of kayaking. With friends Rob Lesser, John Wasson, Royal Robbins, and Reg Lake, they pushed each other to the limit as they notched the first recorded kayak descents of three iconic California rivers in a challenge they dubbed "The Triple Crown." The odyssey involved carrying kayaks on their backs over a 12,000-foot ridge, then paddling through rapids so steep that at times they used ropes to lower themselves down cliffs while sitting in the kayaks. On the Yuba River, Tompkins was out of his kayak and got caught in a whirlpool that sucked him deep down, spit him up, then sucked him down again. Lesser paddled upriver and snuck the nose of his kayak on the edge of the whirl, allowing Tompkins to grab the kayak. "It was this slow killing machine that basically kept rotating back and forth," said Lesser. "He knew he could die in there and his reaction was, *Don't tell Susie*."

Traveling into the wild was not a hobby, it was an essential part of Tompkins. He needed the wild to escape the confines of routine and the stress of running what was now a $15-million-a-year business. Whether working remotely from bad dialups in Tibet, the Swiss Alps, or Borneo, he called his key team members with ideas at all hours.

"Doug wouldn't just come back from a trip; he came back with 50,000 new ideas, and new directions, and new things that had to be implemented," said Tom Moncho, his executive assistant. "Every Wednesday, we met at eight o'clock in this conference room, and then would come out bleary-eyed at, say, six o'clock. Some of [the ideas he offered] were not ready for consumption until they had been polished up a bit. He put things so bluntly, so if you went out and said to everyone what his ideas were, they might be pissed."

Doug curtly said "No" to executives who asked to fly first class—*a total waste of money*, he proclaimed. At his company, everyone flew coach. When the designer Tamotsu Yagi asked for a reserved parking spot near the door of the main offices, Tompkins spit out a quick "No" and explained that no one in the company received such perks. Tamotsu then asked why, in that case, did Doug always park next to the front door? Doug shot back, "Because I'm the first one to work every day."

When Yagi asked for a translator to help him understand English, Tompkins blanched. Translators, he declared, were a worthless use of time. Instead, he inaugurated a language training program, including Japanese language classes. The employee who advanced most in their understanding of Japanese every year was given a free trip to Tokyo. For two.

"The whole company is on a first-name basis," said Doug to an interviewer. "The telephone registry we have on our desks—they're all alphabetical by first names, because people don't know the last names. Someone sent to me a directory that had all the last names. It was completely useless and had to be sent back to be completely redone."

At Esprit headquarters, Doug allotted himself a small space—he wasn't one to seek shelter behind a desk. His office was off the main passageway to the graphics department. "I want to be totally attached to the creative side," he said. "He was always going into design meetings and looking at prototypes laid out on these big tables," said Ridgeway, a climbing partner and documentary movie producer. "It was all open office; there were no walls higher than four feet. There was a sign that said, 'Commit, and then figure it out.' I told him I liked that saying and he said, 'Oh, that's not mine; I got that from Napoleon.'" His other favorite saying was "Life Is Entertainment, Survival Is a Game."

After commandeering choice shelf space in leading department stores in New York, Dallas, Chicago, and Los Angeles, Doug focused on overall branding and merchandising while Jane and Allen concentrated on clothing design and sales. "He was going to Tokyo early on; it was still quite a mysterious, a different place. And anybody from the West lived in this strange bubble and he loved that," said Deyan Sudjic, who worked with Doug and later became the director of the London Design Museum. "The idea of doing a store-within-a-store—Japanese retailing was exactly that model."

Essential to Esprit de Corp's success were the unique fabrics and mix of bright colors. Every month, the production cycle required that Susie order products from Asia. Japan provided the highest quality, so she ordered fabric from Tokyo by the kilometer. As the monthly deadline approached, Susie walked the office wearing colorful knitted gloves she had snagged in a New Delhi market. The colored gloves became her muse, the fingers her palette. "Should it be these two? These three? She would take a long time to make that choice," said April Stark, her assistant at the time. "She had to decide exactly what three different colors she was going to dye these fabrics she was going to use to weave the plaid. . . . To be the color guru was a premier job. We had a colorful business and that was one of Susie's geniuses."

Susie traveled the world investigating colors. A hotdog vendor's awning stoked a bolt of inspiration. Susie copied the pattern, swashed it onto dresses, and had a three-year run. She found inspiration at museum exhibits and nightclubs and spent five months a year traveling. Her perusal of antique markets, garage sales, and Tokyo boutiques fostered original color combinations. "We'd always be deciding up to the last moment, so when we decided on the colors, I would print them up and rush to the San Francisco airport," explained Susie. "We'd ask stewardesses to bring the samples to Hong Kong, and when they arrived there would be someone waiting to pick them up on the other end."

While designing a dozen different tags for her burgeoning collections, Susie looked at the different brands—Plain Jane, Sweet Baby Jane, Rose Hips, Cecily Knits—and decided to simplify the labels. She'd use the name Esprit de Corp on every tag. A single unified name. "I came back from Japan where I was working and told Doug. He was mortified—said it would be the end of our company. He *hated* the idea of a generic label." Within weeks Doug recognized not only the brilliance of Susie's idea but also the urgency.

As Esprit de Corp boomed, costs rose. Manufacturing in San Francisco allowed for flexibility to adjust orders up to the last minute, and delivery was immediate, but margins were tight, so Tompkins broadened his horizons. What if he took the manufacturing out of the United States? Why pay $1.60 per hour for manual labor in San Francisco if he could pay ten percent of that in Asia?

Tompkins expanded his operation in Hong Kong. Ying was hardworking and motivated and became Esprit de Corp's principal Asian partner, a key owner of Esprit International, and a cog in global Esprit operations. In Hong Kong, Susie took over the design of sweaters. She was "busting her tail and living in the Far East for weeks and weeks at a time, being away from the kids," said Doug, who managed marketing and image-making.

Doug explored production centers throughout the world. He sought to move production from San Francisco to India, but it didn't work out. Esprit couldn't get the high quality he required. When the San Francisco workforce (most of whom worked at the Great Chinese American Sewing Company) got wind of the international ploy, they picketed the Esprit de Corp offices at 900 Minnesota Street. Doug and Susie told employees they were not leaving Chinatown to save money but were leaving to get better sewing.

Doug—who never shied from controversy—stuck his mouth in the middle of the Chinatown–Esprit labor dispute when he met with union officials and admitted he was leaving to cut costs. "Doug said that we were moving overseas because we couldn't afford to pay the union wages," said Henry Gruchacz, one of his lieutenants. "You can't do that. If you do, you lose the fight with the National Labor Relations Board and you pay a huge fine. But Doug did; he just stated that union wages were too much to pay."

Schwartz, the top salesman, sensed trouble. "You can't put hundreds of people out of business because you're moving abroad. You go to them, tell them 'I'm gradually moving my production.' You do it slowly, you let people get prepared for their next life. Not him."

Union protesters swarmed the Esprit de Corp offices. Tompkins had them arrested. Union organizers tried to unite "the spirit of the people" into a union, so Tompkins locked workers out. Gerda Kainz, a German seamstress with an encyclopedic knowledge of clothing production and design, knew Tompkins was about to go overboard. "Doug called a meeting at Great Chinese American Sewing and told them there would be no more GCA if the Union would come in. That didn't go down well with them, and as a result I had the tires of my car slit right after that meeting."

Union organizers plastered San Francisco with a "Wanted" poster featuring a bushy-eyebrowed, mustachioed Doug Tompkins talking on the phone. The poster listed four charges against Tompkins, including "locking out 135 unorganized garment workers, primarily non-English speaking immigrant women" and "ordering his truck drivers to drive through the picket lines injuring several workers." The poster urged San Franciscans to be on the lookout for a bearded man "last seen driving away on a Triumph motorcycle headed towards Esprit de Corp." The suspect was said to be wearing brown boots with "the fronts off like a brother from the street."

On January 31, 1976—the first day of the Chinese New Year celebrating the Dragon—saboteurs poured gasoline on the roof of the Esprit de Corp central offices in San Francisco. It was the third time that a suspicious fire had broken out at Esprit, including one blaze that caused over $100,000 in damages. This fire blazed longer than the previous. As the burning roof collapsed into the Esprit HQ, the company's management team sat for a long-awaited gourmet meal at Chez Panisse, Alice Waters's restaurant across the Bay in Berkeley that was still wait-listed—even five years after its 1971 opening. Their meal was interrupted by a waiter who told them that a phone call had come in with urgent news: there was a fire at headquarters. But it had taken months to secure a coveted table and the team had already ordered, so the Esprit executive team waited. When the food arrived, they wolfed it down and impatiently sped back to San Francisco. Doug's assistant Tom Moncho never forgot the view that night from the Bay Bridge looking into San Francisco. Enormous tongues of flames lit the night sky. Their Esprit de Corp was ablaze.

Gruchacz, Doug's right-hand man, who had not gone to the dinner, was at his home, just several blocks from the office. He had already heard sirens and smelled smoke when he received a call. "They said, 'Henry! Go down to the factory. It's burning! The union burned our factory down,'" said Gruchacz, who described rushing down Minnesota Street, the sky alit by flames. The heat inside the building was so intense that some seventy firefighters fought to control the inferno. Fire

Department investigators never proved arson, but a firefighter told the *San Francisco Chronicle* “that’s not the way buildings normally burn.”

Sweeney, the furniture craftsman, took a call from the property manager, Rex Wood, who told him there was a fire and that he needed to secure his woodworking studio. When they arrived, the building was nearly gone and exposed to the street. “Rex had a gun. And he gives me the gun. He holds the flashlight,” recounted Sweeney. “And we’re walking through this darkened space, water dripping everywhere, charred things about us. And we were walking toward my shop to make sure the shop was okay. And I think, *Whoops, I don’t know how to use a gun. What if I shoot Rex by mistake?*”

Tompkins was in Germany sleeping on the couch of Jürgen Friedrich, the Swiss owner of Esprit’s European operations, when he received a phone call from his assistant, saying his company was on fire. “I’d never seen him crying,” said Friedrich. “But this night he was crying like a baby because his life’s work was gone. So too his priceless and rare collection of early American quilts. But it was only for an hour. Then he picked up the phone and started firing off directives. He was calling all around the world. The phone bill for that single day was four hundred Deutsche marks.”

As Tompkins boarded the flight to San Francisco, a central question gnawed at him—the computer files. Were they safe? Tompkins knew that the building (now ashes) had been a winemaking factory and that the backup computer discs were stored in a vault built a century earlier by vintners. Tompkins had added a steel door, providing a second level of protection for a thick, fireproof safe holding IBM magnetic tapes. Those tapes, fragile and sensitive to heat, were the DNA of Esprit.

Gruchacz entered the safe and found the discs covered with smoke and oily soot. “I knew that we were in deep shit if we didn’t have those discs,” he explained. “All of the sales and all of the accountings was on those damn discs. . . . I took them down to IBM. All the data was there. We still had a company.”

Had the discs been destroyed, "we would have probably just folded because we would have been in such radical turmoil," said Tompkins. "We didn't have enough insurance."

Gruchacz and a clutch of top personnel went to the airport and picked up Doug and drove him to the scene of the fire. "He was not hysterical. He just looked at the wreckage," said one employee.

Outside the shell of their corporate headquarters, Doug joined Susie. "We were sitting on the curb," she said, "looking at the smoldering ashes and he didn't miss a beat. 'We'll just rebuild it even better' and I was kind of stunned. He didn't pause."

Chapter 5

Esprit de Corp

I was never attracted to fashion in and of itself. What interested me was not design per se, but the lifestyle that is attached to products. The products are not really all that important; it is the aura around them that counts in many cases, and that is what interests me. Ideas start with a view and reflection of society and then are articulated through the product. You breathe life into the product by weaving a personality around it.

—DOUG TOMPKINS

Tompkins consolidated his vision and control of the company following the fire. An ad in the national press announced that Esprit was "Down but not Out." Working around the clock in Doug and Susie's living room, employees rebuilt shipping lists, reassured clients, and embraced the challenge as industry underdogs. One week after the fire, Esprit was shipping. "It was like the earthquake and fire of San Francisco for us, but disasters and calamities have a way of uniting people in a common cause," said Tompkins. "It was a rallying point and people really responded to that and understood. . . . That was a company time of group spirit—esprit de corp, if you will."

After the fire Tompkins saw a chance to take full control of ownership and buy out the original partners, Schwartz, Dwelle, and Tise. It was a messy affair. "In order for me to get my equity, they made me release $375,000 in commission. They held it over my head," said Schwartz, who owned 20 percent of the company and was head of sales. "I got screwed out of that. And I got screwed out of the whole company, period."

Doug and Susie forced out a founding partner, Jane "Plain Jane" Tise, the visionary whose lines of blouses and dresses led the company's initial flair. After she reached a deal with Doug and Susie, "Jane had the greatest line about this whole thing that I ever heard," Allen recalled. "She said, 'After they get finished fucking everybody else, then it will be fun to watch them fuck each other.'"

"Allen was expensive, he'd take clothes back all the time and that cost us," retorted Bill Evans, an Esprit de Corp executive. "He was overpaid."

Schwartz disagreed. "I had contacts with every single department store and chain in the United States. I took that company from a dress company to eight different divisions, while this guy Tompkins was canoeing and mountain climbing."

Tompkins noticed that as soon as Jane left, Susie bloomed. "It was a huge psychological burden off of Susie," he remarked, "because Jane was very domineering in her design ideas. And Susie was kind of intimidated and really didn't go up against Jane's ideas. But when Jane left, Susie took over the design direction and that was like a rebirth. She flourished. It became evident shortly thereafter that Susie, instead of Jane, was the real hidden talent."

As they rebuilt Esprit, Doug recognized his primary challenge was not to sell clothes. He was selling an image, fulfilling unconscious wants, designing dreams. Doug loved the challenge: understand the needs, pinpoint the cultural zeitgeist, market the product. Sales would follow. He was also clear about audiences that were of no interest to Esprit. "The yuppies are not our market segment," he said. "They don't have enough sense of humor. They're too cutthroat to laugh at themselves."

By the late 1970s, Tompkins had presciently diagnosed the '80s as an age of narcissism and pursuit of the good life. Esprit represented a Northern California–fit lifestyle that included gays, clusters of blond boys, and a burning sexuality bordering on a gleeful celebration of group sex. In a wild moment post-pill and pre-AIDS, Esprit become a cultural landmark for a lifestyle as utterly delicious and desirable as it was unattainable. The Esprit world was inhabited by blonde women wrapped in a style that indicated they were free, available, and healthy, akin to a vegan version of *Baywatch*.

Working with ace furniture maker Sweeney, Doug envisioned the new Esprit de Corp offices as a showcase for design. Together the designed elegant desks, hand-crafted chairs, and lighting befitting an art gallery. When finished, the entrance to the new Esprit de Corp offices was breathtaking. The employee cafeteria was designed as chic as a snack bar at the MOMA, and what remained (after the fire) of his Amish quilt collection now hung on the walls, the width of each picture frame decided by Doug. "There were exercise classes. And you could go to cultural events," said Dan Imhoff, an employee who became very close with Doug and eventually married his daughter Quincey. "Maybe you didn't make quite as much money as you possibly could, but you were all in. There were these amazing parties and trips and you could go to Tahoe and ski for the weekend. It was like college meets work."

For Doug the goal was to be the best overall clothing company in America. He knew that Calvin Klein was more sexy and Liz Claiborne more profitable, but he thought of the clothing business as a kind of multisport proving ground where varied skills were needed. Esprit was not top in any one discipline but, as Doug bragged, "We're good at data processing, we're good in finance, we're good in design, we're good in image. Maybe we don't have the best department in our industry in any one category but it's #3 here, #5 there, and so forth. When it adds up, it comes out that maybe we average out to be the best." He called the overall strategy his "decathlon concept."

Yvon was running Patagonia in a similar fashion down the coast in Ventura, California, where the company pioneered onsite childcare, used recycled paper for its catalogs, and donated millions to protect old-growth forests and fought against the damming of free-flowing rivers. Yvon and Malinda saw clear benefits to staffing their offices with long-haired surfers, hawk-loving secretaries, and a wild tribe of passionate outdoor adventurers. These were the people who would find a way when Patagonia boldly migrated from cotton produced with chemicals and insecticides to organically produced cotton. For a clothing company, it was a revolutionary gamble, and Patagonia's success, plus its willingness to funnel donations to direct-action activists (including the legal defense of Earth First! leader Dave Foreman) gave the company deep credibility with a generation of up-and-coming environmental activists. A graduate student named Mark Capelli pushed a plan to save the steelhead in the Ventura River; Malinda and Yvon found room for Mark in the cramped offices and realized that having an environment officer spread the green message throughout the young staff.

At both Patagonia and Esprit, many employees were self-confident individuals who might dye their hair blue one day and bring their dog to work the next. Like Paul Hawken, who created the gardening gear outfit Smith & Hawken, and Stewart Brand, founder of The Whole Earth Catalog, these men forged companies with the values of '60s rebellion and counterculture. Led by lifelong rule breakers, these companies were nimble when opportunities arrived and unfazed when thumbing their nose at conventional wisdom, which they smugly decried as lazy thinking. While Silicon Valley tech companies later took credit for founding "the corporate campus," it was Doug and Susie at Esprit and Malinda and Yvon at Patagonia Inc. who launched in-house organic salad bars, created a family-friendly workspace, and paid their workers to volunteer with AIDS patients.

I had no idea who he was. There was no internet, so there was no way to find out about him, so I didn't bother. We sat down, and he got very belligerent right off the bat. "So, you're this wise-guy designer I've heard so much about. You just rolled into town." I said, "Okay, let's get over that. What is it you want?" He told me the story of where they were with Esprit, he and his wife had this business, they were doing $10 million a year.

The first assignment is that we went at the name Esprit de Corp. I said the name is too long. He said, "Yeah, we should just cut it down to Esprit." He said, "How much time do you need to redesign it?" I said, about a month. I went off, put some thought into the whole thing, and came up with the three-bar E, developed an alphabet for him. This was for the transition of the other brands; at that time he had eight or nine brands. So, I go in, it was only him. It was very rare for me to make a presentation to only one person. Typically, it's a board. He was very relaxed, and I made my presentation. He looked at me, and said, "This is it?" I said, "Yeah." He said, "Well, umm, I don't think you worked hard enough. I think you need to go back and spend more time." I said, "Really?"

I gathered everything up. He said, "How much time do you need?" I said, "Gimme another month." So, another month goes by. In the meantime, I do nothing. I didn't touch it. I call him up, say, "I'm ready to be with you again." He says, "Great." I go into his office, unravel the same presentation I had done a month earlier. He realized it instantly. He said, "What the fuck are you doing? I asked you to bring me some more ideas!" I said, "There is no other idea. This is it. You're crazy if you don't do this." He was flabbergasted. He looked at me, and because he realized what was going down—he realized my ego was on the line—he realized I was absolutely certain. If he asked me to do it again, I would have said no. I would have said, "Get somebody else to do it." And with a smile he said, "Well, if you're wrong, I'm going to kill you."

—JOHN CASADO

In his quest for perfection, "Chairman Doug," as some employees called him, searched worldwide for the ideal photographer to shoot his Esprit catalog. International photographers, including Roberto Carra, the head of graphic design for *Vogue* Milan, were brought in to prepare the imaging for Esprit. But leading fashion photographers, one after another, struck out. They battled with Doug over aesthetics and clashed with his ego. Tompkins enjoyed the conflict and, not surprisingly for a devotee of fencing, appeared to gear up for the next round of jousting. To recruit Oliviero Toscani, the son of a celebrated Italian photojournalist, Tompkins learned Italian. Toscani infused fashion with an editorial attitude. It was exactly what Tompkins was looking for.

In January 1980, Tompkins and Toscani scheduled dates for the new catalog shoot. They wanted a new image, a new feel, an air of innovation for the '80s. "He was kind of wandering around, and the question was in the air—Well, what are we going to shoot?" said Dwelle. "I suggested, 'Let's go in the conference room, and stay here until we can say, *Why this catalog? What is it trying to say, and to whom?*' And we sat there for a couple of hours. Finally, Doug said, 'What we're trying to convey is not the product, but the collection, the look, the way of being.'"

When Toscani agreed to work with Esprit de Corp, the bond with Doug was immediate, both in lifestyle and aesthetic taste. Oliviero was as ego driven and brash as Doug. Together they delighted in snubbing the fashion industry. At the top of the mountain as measured by market acceptance and street credibility, Esprit de Corp announced it would no longer hire professional models. Tompkins and Toscani created one of the first "real people" storytelling campaigns. Esprit de Corp employees, then customers, were recruited from stores around the world and featured in the company's advertising campaigns with quirky phrases, including "I'm looking for a sensitive man who is also a sushi chef" and "There are two things about me that are small: my feet and my bladder."

"Toscani was very helpful in teaching me the craft of image making," wrote Tompkins. "I got into that circle of people and started to learn the process and the discipline. The whole industry was looking at what we were doing. All great architects, designers, artists, brands, companies establish a style and change it with slow, careful modifications, always keeping all their customers following along with them. Lifestyle changes are slow; they do not jump from moment to moment."

The Esprit catalog shoot became an epic event. Thousands of customers lined up for the castings. Locations were scouted worldwide. "[Doug] had an inner power to him that reached out, and you felt like you were in the presence of somebody that had his fingers on a whole lot of things," said Helie Robertson, the photo coordinator for the catalog. "He had good taste. Energy. He could climb up the side of Yosemite. He could commandeer a plane. He could drive fast and have fun. He made things happen."

Esprit grew famous as a maverick company, as the cutting edge of teen fashion. Job applications and fan mail poured in. When Esprit went to Stanford to recruit college seniors to consider a career at the company, they were flooded with applicants—from recruiters working for the other companies at the job fair! There was no doubt, Esprit had pizzazz. At a New York shoe show, Doug had his staff serving sushi to the crowds, while also talking up the company. "I'd rather have our people serving the sushi at a show in New York than a caterer. It's more interesting. . . . They're not learning to be waiters and waitresses, but they're learning how to interact and impress our clients," crowed Doug. "We're showing that even our employees represent the lifestyle of the products that we make."

But the attention to tiny detail also led Esprit employees to wonder if Tompkins had lost his mind. When Tompkins needed a unique-sized napkin for Caffe Esprit, the restaurant he planned to open near the Esprit San Francisco outlet store, he dispatched an employee on a trip to investigate napkins. Eight months after embarking, the napkin researcher returned and Doug announced that his napkin, which he considered a work of art, would measure exactly sixteen inches by sixteen inches.

The doggy bags at Caffe Esprit went through such a rigorous design and vetting process that they became an underground fashion sensation, as coveted accessories, and even used as trendy pocketbooks. Every sales slip carried a multicolored swirl created by Tamotsu Yagi, Doug's favorite Japanese graphic artist and designer.

The Esprit company newsletter was playful and published a weekly mixture of gossip, helpful tips, and satire. A tongue-in-cheek "Me Generation" column by Lee Rosenberg offered a lecture series for employees explaining "Guilt without Sex" plus special tips on "How to Convert Your Family Room into a Garage."

Esprit became so popular that market research companies identified the company as the definition of young, upwardly mobile, independent women. Marketers defined this new tribe as "Esprit Teens" and the suitors lined up. Helen Gurley Brown, the legendary editor-in-chief of *Cosmopolitan* magazine (circulation at the time three million copies a month), noticed the splash that was Esprit and made a face-to-face pitch for a share of the juicy advertising budget. She came to Doug, her heels held in her hands to protect his soft balsa wood floors—and gently urged an alliance. Tompkins mocked the *Cosmo* cover girls. "I find them sex-object types, flaunting their cleavage. They've got these phony-looking hairdos, and they've got a lot of makeup," he told a stunned Gurley Brown. "That's not the kind of image that I'd like to portray, so I don't think that so many of our customers are reading that magazine." Tompkins popped a mint into his mouth and concluded, "I guess it's just a matter of taste." Maureen Orth, who was in the room penning a profile of Tompkins for *GQ*, described Doug's rule over Esprit as a "benign dictatorship" and "not everybody's cup of ginseng."

Young Esprit employees, however, understood Tompkins and thrived at a workplace they called "Little Utopia" and "Camp Esprit." Far from being aloof, Doug was approachable. "It comes down to the hard work, rolling up your sleeves, and being involved in the nitty gritty of daily action so you really know what's going on," said Doug. "All these guys who run large organizations become detached. Sort of like the President of the United States who seems so often to be detached from what's happening on the streets."

Esprit also sought to build bridges into the local community. "Esprit had very good relations with the neighbors," said Bill Evans, a manager at Esprit. "This was a warehouse building that was in a black, residential neighborhood, working-class people. Doug made a deal with all the people living nearby, and planted trees. Drew, the plant manager, took out a piece of the sidewalk, planted trees, and painted houses immediately adjacent to the Esprit building so they looked more presentable."

When a house near the Esprit offices burned to the ground, Tompkins leapt at the opportunity and bought the lot. First he razed the buildings, then the experiments began. The city lot was Doug's oversized petri dish. How much park could he add to a single lot? Do you put in tire swings or a zip line? If pedestrians want to rest on a bench, won't they also prefer shade?

Two years later the park was completed with a fishpond and a picnic area. It was designed as a lunch spot for both Esprit employees and anyone from the street. On the other side of the Esprit offices, Tompkins bought the entire block, which had been a metal galvanizing yard. He removed the soil, because it was contaminated. And after bringing in new soil, he flew to Oregon. In his plane he circled a farm with trees for sale—large, thirty-foot-high trees that required special flatbed trucks to transport the root-ball some 900 miles south. He bought several truckloads and thereby seeded his tiny urban nature escape with a jogging trail complete with redwood trees.

Years before Google cofounder Sergey Brin was born in a Moscow public hospital, Doug and Susie Tompkins, the hippest of the hip in San Francisco, nourished a workplace so employee-centric that some called it a cult. In addition to free Italian classes, kayak excursions, and company-wide Halloween parties, Tompkins pressured his workers to leave town. To go away. To escape. "Frankly, I don't want to underwrite anything that's just a vacation," he said. "Everybody can take a vacation lying on the beach in Hawaii, but how many people will really go rafting in the Himalayas? It's a win-win situation. The individual will heighten their sensibilities about being alive, and if they are alive and more dynamic, the by-product goes to their organization."

"I was hired to be the liaison between the sales department and the design department because I had owned a pretty successful and hip boutique in San Francisco," said Helie Robertson, who worked at Esprit for years. "They always made it real nice for you. There was a house with a swimming pool. You got all your food free. You would spend the whole day working. Then you'd eat out with everybody. On the weekends, you'd go do something fun. It was always a great atmosphere for making sure that you had fun time. They didn't expect you to keep your head in the business the whole time, because I think they realized that it energized us to feed our outside interests."

Esprit intercompany sexual affairs were deemed inevitable, as employees from all levels of the company mixed and met at the Esprit tennis court, on the Esprit volleyball team, or sleeping at the Esprit chalet along the romantic wooded shore of Lake Tahoe. "I think Esprit was pleasant to us," said Dwelle, who lost his wife in the sexual shuffle that was Esprit. "It was too much opportunity, too much success, too much money, too seductive. We lost track of each other and we lost track of our kids. You know, we lost track of anything except following those tracks. Neither of us did drugs, but it was like a drug."

As a prize for top division employees, in 1982, Esprit offered a whitewater rafting trip to the Biobío River in Chile, among the most rugged rivers in the world. The Biobío was the same river that Tompkins crossed on his Fun Hog trip fifteen years earlier, and now he brought his family. Doug, Susie, Quincey, and Summer all went on the rafting trip.

The Esprit team flew to Santiago and were met by guides from Sobek, a pioneering whitewater rafting expedition company working in a Chile that was still under the dictatorship of army general Augusto

Pinochet. The military government was brutal. Armed forces members assassinated an estimated 3,000 civilians using a technique known as "detain and disappear" in which activists were arrested, tortured, murdered, and their bodies dumped from helicopters into the ocean. Given a popular insurrection and weekly street riots against the dictatorship, tourism was nonexistent. Pinochet's secret police recorded the Esprit team's visit in their files: Señor Tompkins, who had gotten in trouble with the stolen BMW motorcycles, had visited Chile yet again.

To raft the Biobío, the Esprit team left Santiago and headed south through a long valley filled with orchards, wineries, and fields of wheat. They passed near Colonia Dignidad, a German compound that offered roadside dinner and fresh-baked cookies for weary travelers. Situated on 40,000 arable acres, Colonia Dignidad was a mysterious spot cloaked in rumors and legends. Stories abounded that surviving rulers of the Third Reich were hiding inside the compound. No one ever spotted a war criminal, but it was a common sight to see rows of uniformed men in paramilitary training exercises. Massive arms caches hidden underground only added to the doubts about activities beyond dairy farming. Male and female subjects were kept under harsh discipline and forced to reside in separate living quarters. Beyond the milk cooperative and a public health clinic, the Germans were also running a classified intelligence gathering operation on behest of extremist right-wing movements seeking to crush leftist rebel groups in South America. General Pinochet sent a regular stream of prisoners and visitors to Colonia Dignidad, including mothers to be tortured and his own children to enjoy fresh air and a secure location to frolic in the countryside. The Esprit crew didn't stop for cookies.

Arriving in the Biobío region, the Esprit team boarded a locomotive up the valley. The steam-powered train puffed into the mountains. Like pioneers from the 1850s, the Esprit crew purchased pots and pans at a rural outpost where cheap wine came in five-liter glass jugs. But hours after they arrived at their first riverbank launch spot, a stomach virus erupted. Employees doubled over in pain, vomiting, half-paralyzed, barely able to walk. Instead of the promised hot weather, the temperature dipped to near freezing. One Esprit employee slipped into the icy waters of the Biobío River and had to be rescued. "Nobody got hurt, but it was a nasty swim for someone with a terrible, cheap wine hangover," said Dave Shore, one of the guides.

Yet for the Tompkins family it was a rare moment of togetherness. By day they floated the headwaters of the Biobío and at night camped at "Santiago's Wall"—home of a toothless cowboy, dressed in a black poncho, who brought a goat or sheep for a riverbank BBQ. Romances between the grizzled river guides and the Esprit women on the trip led to numerous affairs and, within a year, a trio of marriages.

Doug descended the river in his kayak, exploring with his expedition buddy Royal Robbins. For Susie and the kids, who played with Royal's daughter Tamara, the trip was a break from the chaos of separate travel and increasingly disconnected lives. Susie was now spending months a year in Hong Kong, often taking her young daughters. Doug stayed in San Francisco, then escaped on adventure excursions that consumed several months a year. The family was fragmented, but the cracks were invisible to all but those in the inner circle of Esprit.

In 1983, Esprit sales topped $1 million a day. Doug and Susie were San Francisco business celebrities, and their guest house was a crash pad where neighbor Steve Jobs or comedian Robin Williams might be found. "Robin Williams drove Doug crazy. Doug thought he was way, way too hyper," laughed Rick Ridgeway. "Robin was a runner. He was staying at the house, and Doug was so tired of listening to him talk that Doug said, 'Let's go for a run.' Doug ran Robin all over the city just to make him exhausted. Doug said that night Robin was so quiet, it was the best night he had with Robin Williams."

Once a year Doug hosted a day-long "Pepper Party." Tompkins, an accomplished cook who mastered the magic of marinade and slow-roasting red peppers, held sway over a crowd that included his heroes in climbing and kayaking together with artists of all stripes. To guarantee a collection of a dozen mad geniuses, Tompkins printed different versions of the invitation, each one custom formatted to inform a VIP that he or she was guest of honor. Even when the celebs understood the trick, they found reason to attend. "Yvon Chouinard is there, and Diahann Carroll, the singer," said Rob Lesser, the kayak photographer and filmmaker. "The problem was you didn't know who the hell they were, and they probably were captains of industry. But you would not find Doug associating with people who were blowhards or full of themselves. You found Doug with people who were interesting, and who stimulated him. His mind was so damn quick that he could see through you in a second. If he was in a conversation with somebody who was wasting his time, he was adept—without being offensive—and would melt away."

Inside the main house, his art collection included paintings by Botero and a tryptic by Francis Bacon. "When Doug was in his art acquisition phase, I was often with him, in London, where we were going to galleries," said his friend Peter Buckley. "He would, first, look at the entire repertoire of that artist. Then he'd get the stories. And then he'd isolate; he'd say, 'This is the one. This is the masterpiece.' And it was a sort of methodology he took to every artist that he was looking at."

When Ridgeway slept over with his wife, Jennifer, Tompkins told them to be careful and not touch the painting above the headboard. Rick thought, "Of course, he doesn't want some oily hair or head to leave a scratch." And then Doug said, "Because if you shift it at *all* the alarm goes off in the downtown police station." The next time Ridgeway viewed the artwork it was on display in Paris, at the Centre Pompidou.

Doug Tompkins had become the bad-boy prince of American retail. Esprit's colorful clothes exploded off the shelves. Magazine covers framed his face with banner headlines like "The Word According to Doug." Esprit was among the hottest global brands, and the DNA was all Doug. "The ethos he was projecting—and after a little bit of practice, he was able to articulate it—was 'we're in the business of creating image,'" said Dwelle, his business partner. "You have to live in that mind frame. You have to project an image, the 'me image,' which is clean and sociable and exciting and sexy and dynamic, because you can't very well project that image in your work if you don't internalize it."

Tompkins considered the title CEO far too corporate for himself, too straight. So he had his business card—after numerous prototypes—printed with the title "Image Director," which justified his jaunts around the world, learning Italian, and buying airplanes. For nine months a year he was "Image Director" and for the other three months he experimented with what he called "my own MBA," which Yvon jokingly called "Management by Absence." Their wild adventures never failed to invigorate new thinking and provoked deeper responsibility and independence in their senior staff. Neither man had any desire to take their companies public and risk sacrificing intrinsic value for what they mocked as "stock value." With virtually no corporate debt and little outside financing, both Patagonia and Esprit were profoundly designed to cater to the leadership of their founders. Leaving the office for months at a time was part of their routine. "It didn't bother any of us," said Robertson, who worked at Esprit during monthslong absences by boss Doug. "It might have even been a relief at times."

Exploring the wild corners of the planet also allowed Tompkins to scout retail opportunities. For nearly two years he negotiated with the People's Republic of China in a bid to become the first American clothing company to sell retail clothing in the PRC. Given the restrictions on clothing imports into the US, Tompkins obtained special tariff exemptions to import tens of millions of dollars of clothes and accessories from Hong Kong. In the entire US Doug held the largest single import quota from Hong Kong. He'd seen the market early, grabbed a share of the world's top-quality and low-priced manufacturing, and was now selling those clothes to some 7,000 stores worldwide. His empire was based on inexpensive, mass-produced clothing. When the catalog was printed in Pennsylvania, the final tally was 1.5 million copies, which, as the company newsletter bragged, was the equivalent of "22 railroad cars filled with raw paper just to print that all up."

Tompkins was able to abandon his role at Esprit for weeks at a time only because Esprit was a privately held corporation. Loathe to dilute ownership or power, Doug and Susie kept Esprit resolutely in their private hands and thus immune from shareholder revolts, stock price concerns, or issues of management control. Even as annual revenue neared $600 million and sales expanded to dozens of countries, Doug would often escape from the San Francisco corporate offices for three-week-long expeditions to Bhutan, or for a month to notch the first descents of whitewater rivers around the globe.

Doug and I decided to run the south fork of the Maipu River above Santiago [in Chile]. It had never been run before. We got dropped off, and we are running this river and we came across one area that had a sharp corner. We could hear a roar on the other side, so when you are running a new river you've got to get out and scout it. I stayed in my boat and Doug got out and walked up a knoll to look around the corner, and while he is looking around here come two soldiers. They stick their guns in his back and demand to see his passport. He said, "What the hell you talking about? I don't have a passport, I am a kayaker. Look at me? I got shorts on!" They are calling in, back to headquarters, and he takes off and comes running down the hill and jumps in his boat without telling me anything, and zips around the corner. I am freaking out. I am thinking, Oh my God, I don't even know what is around the corner, these guys could shoot us, *so I go around the corner and I am so freaked out that I am sitting way up stiff in my boat and, sure enough, I hit this reversal and I go over and I am upside down. And I am thinking,* How can I stay upside down as long as I can? I don't want to come up and get shot. *Later that night we found that we were at [dictator] Pinochet's summer estate, which we didn't know anything about.*

—YVON CHOUINARD

Tompkins no longer needed the ego boost of a Ferrari. Instead, he strapped a bright red plastic kayak to the rooftop of a beat-up, black station wagon and bombed around the Bay Area. People gawked. A kayak atop a car in the city was strange. People stopped and questioned him. "I had one kid sitting around the car until I showed up. He said he had been waiting there fifteen minutes to find out about this thing. This guy was a surfer, in Monterey, and he had to know about this weird-shaped thing. They know it is a boat, but where does it go?"

On his trips, Doug avoided five-star hotels. He preferred to sleep on the couch of a friend with his sleeping bag, always the first item crammed into his travel bag. "I can have a private jet, but I can't imagine it," he said. "I poke along in my little, single-engine propeller airplane and land on beaches."

Tompkins's appetite to learn more about design grew. He read book after book about both the cutting edge of design and architecture, and then recruited superstars Joe D'Urso, Ettore Sottsass, and Shiro Kuramata to build Esprit stores in Manhattan, Milan, and Tokyo. For those in the know, it was like having Maradona, Messi, and Pele on the same soccer team. "Doug loved the experience of working with designers who were going to shape the places he worked and lived in," said Peter Buckley. "In Japan he got Kuramata, who designed the Singapore and Hong Kong stores for Esprit, to do a house, which is where Doug stayed when he was in Japan. He loved those experiences."

After hiring John Casado to redesign the company's logo, Doug invited him over to his house on Lombard Street. "It was this fabulous green shingled house on Russian Hill, it went from block to block," recalled Casado. He immediately noticed the white plank floors, and the very modern look. Doug was busy doing something in the kitchen, and told him to hang his coat up in the closet. It was a walk-in closet, and as Casado hung up his jacket, he had a realization:

it wasn't a closet at all, but a library. "Every book on the wall was about design, or style, or architecture. I was flabbergasted, because I have done pretty extensive reading on design. I recognized he had very esoteric books on design. When I walked out, I said, 'I give up, are you a *closet designer*? What is all this?' Doug replied, 'No, I just love design.'" That calmed Casado down. He realized he was working with a client that he didn't have to educate: someone who would appreciate what Casado could do for him.

Steve Jobs would later mold the early Apple flagship stores as indirect tribute to the beauty and flourish of the emotional retail experience crafted by Esprit. When Tompkins published *Esprit: The Comprehensive Design Principle*, Jobs ordered his team to buy copies of the six-pound book. "When I was working on the Apple retail store concept design with Steve Jobs in 1999, he wanted to emulate his store on one of the stores within the Esprit book, and he bought fifty copies from me to show his design team. He called it a 'design bible,'" said Yagi. "Douglas Tompkins reminds me of Steve Jobs, or should I say Steve Jobs reminds me of Douglas Tompkins?"

Tompkins was eclectic in his '80s tastes. Perhaps the lack of a defining style for the decade itself allowed him to graze among different flavors. For a few years he tried Italian. He had a love affair with all things Japanese, and then fell (briefly) for the sleek, modern, hi-tech style of the British architect Norman Foster. Sir Norman and Doug were indeed an odd couple—the Lord and the kayaker.

Foster was as revolutionary as Doug. He had upended the London skyline with a building popularly known as The Gherkin, his futuristic curved dome. In a fit of empire building without limits, Tompkins tasked Sir Norman with designing Esprit headquarters as a multimillion-dollar art project, a sculptured city where a whole world could be created. Esprit offices would be surrounded by a soccer field, a daycare center, and a gym. The corporate organic garden was as well designed as the natural lighting for the conference room. South of San Francisco, real estate prices were lower, and redwoods still dotted some of the hills. Tompkins picked a little-known city on the San Francisco Peninsula called Cupertino. Although Tompkins never built his campus there, Steve Jobs did.

Cupertino became the world headquarters for Apple, and Jobs was praised for his visionary thinking. Tompkins, meanwhile, was busy berating Jobs himself. "I used to fight with my friend Steve Jobs, who believed that the Internet, the personal computer, and eventually of course the cell phone were the roads to paradise," he remarked. "These were going to lead us to the Promised Land. I told him they were ruining the world. He'd get huffy about that."

As a forty-three-year-old New York native who'd never finished high school, Tompkins's high-flying life in the mid-1980s was intoxicating. But his love of design and architecture was also wildly expensive. "He was over in Italy building all these monuments to himself. Big store in Milan. Twenty million dollars down the rathole just so he could play with the designers," said Susie, his full-time business partner and, by the mid-1980s, his part-time wife. "Same thing in London with Norman Foster. This is what was going on. And it's just driving the company into the toilet. We couldn't make the US payroll, and he's spending millions of dollars all over the world."

For fellow executives at Esprit, Tompkins's largesse created a dilemma. Clearly the company could not continue without his drive. But could it survive with him at the helm? "In the long run, you have to say, 'God, you worked with all these great architects—Norman Foster, the Japanese, Joe D'Urso—and yet nothing survived. Nothing,'" said Dan Imhoff. "It's sad. When I was in Italy, they moved a wall three inches, at the cost of a million dollars in a building that they didn't even own. It was a leased building. He had lost his mind, in my opinion, and I have a lot of respect for a lot of things that he did, but in some ways, he went over the top. He couldn't ever stop these fanciful things that he would get doing. But, *hey, it was his money to play with.*"

By 1983, Tompkins was afflicted with an overdose of self-confidence. Faith in his unique aesthetic compass was leading him astray. Without fully recognizing the error, Tompkins had let his trademark motto, "No detail is small," mutate into "No detail is too expensive." He custom-ordered steel staircases for his stores, then added chrome bannisters. After picking out his Venetian stone tiles, Doug then became involved in the design of the shipping crates. He collaborated with his mentor Ettore Sottsass, of Milan's postmodern "Memphis Group" design movement, to design the furnishings for Esprit's Milan showroom. One critic described the result as a unique collision of geometry that was "part Bauhaus, part Fisher-Price."

But it was on the corner of La Cienega and Santa Monica boulevards in Los Angeles that Tompkins sought to create a retail showroom like no other—a shrine to design where he could pack underwear into transparent yogurt containers and display towels inside plastic bags designed for a baguette. Tompkins purchased Flippers Roller Boogie Palace, the roller skate disco known as "L.A.'s answer to Studio 54—on wheels."

Flippers was the place to see Gene Simmons in full KISS makeup or Prince in regalia. The Go-Go's, John Cougar, and The Ramones played live shows. Tompkins dreamed that this chic showroom would sweep Los Angeles off its rhinestone heels. He brought to town (or so he imagined) a sexy new scent: Northern California succulent. The Los Angeles location where Jon Voight roller danced with Jane Fonda would, he declared, be reborn as the epicenter of design and youthful clothing. Esprit's Los Angeles showroom oozed the company's ethos that life "is an attitude, not an age." But as many newcomers to The City of Angels discover, faking an attitude is harder than it looks. Tompkins demanded that the concrete floors be painted glossy black, then waxed. They came out smooth, striking, and so slippery that employees and customers regularly crashed to the floor. A contractor working at the L.A. store observed, “You got the feeling that Doug liked to work off a full-scale model.”

While competitors like the Gap spent about $70 per square foot for their properties, the cost of the former roller disco ballooned sixfold to $400 per square foot. “It was to be a masterpiece for him. Everything was overdesigned,” confessed Aldo Cibic, the Italian collaborator who worked on the project. “Like, if you make a house for yourself and you start to say, ‘I want the floor the best of the best. I want the walls the best of the best. The lighting must be the most beautiful. The furniture has to be fantastic.’ ”

The Los Angeles store bombed. “Doug completely ignored the first three rules of retailing: location, location, location,” recalled friend Buckley. Esprit publicly acknowledged that the store makeover, budgeted at $7 million, ended up costing $24 million.

“*The Wall Street Journal* called me for a quote. I guess the banks were worried about Esprit. I said ‘Esprit is a multimedia international work of art,’ ” laughed Dwelle. “Doug was furious. I said, ‘Well, isn’t that the truth?’ He said, ‘Yeah, but you should have told the banks something about the business.’ ”

Seduced by eighteen years of soaring profits, by 1986 Tompkins had missed multiple signs that his company was in trouble. Fashion trends were migrating from flashy colors to a more traditional palette. The US dollar sank, thereby spiking prices charged by his Hong Kong manufacturing operations. Profits plummeted to $8 million annually. In a panic, Tompkins fired a third of his 2,000 US-based workers and cancelled the free ballet tickets and rafting trips for employees. But his deeper problems were closer to home. A turf war was brewing with his wife, Susie. “We’ve been fighting like cats in a bag for fifteen years,” Tompkins told *GQ*. “It’s just a matter of convenience that we remain married.”

Yet the powerful split ran far deeper. For board members and the few in the know, Esprit was cracking under the founding couple’s nonstop warfare. "There's an invisible line that runs right through the design room. You're either on Susie's side or you're on Doug's," said Brielle Johnck, Susie's personal assistant, who quit after becoming "worn down by Susie and Doug's unhappiness."

"Everything that was a problem was my fault," declared Susie. "And so, when we hit the wall financially because we'd overextended and because he was doing all this monument building, it was 'my fault.' I was standing upstairs at our house on Lombard Street, in front of my entire design team, and I heard him say, 'We've got to get her out. She's too old. She doesn't know what she's doing.' That was the night I left. That was the final straw."

Tompkins hardly seemed to notice the tremors his marital instability had on Summer and Quincey and his own contributions to Esprit. He felt restless. He seemed to be impatient. If he died, what would be the inscription on his tombstone? "Nearly a Member of the Olympic Ski Team"? Or "Notable Clothing Salesman"? Hardly the legacy he imagined when as a sixteen-year-old he told an interviewer that he was destined to "go places normal humans don't."

While company insiders knew about his marital misgivings, Tompkins was also doubting everything about corporate America. Tompkins didn't want to play the game anymore. "I found myself caught up in the marketing," he reflected. "I lost track of the larger picture. I was creating desires that weren't there. I was making products that nobody needed."

"He had a sense that business was cool," said his friend Lito. "He was doing it with great enthusiasm for the hell of doing it well. But I never had the sense—ever—that he was trying to make money. He just wanted to succeed because that's what he was doing. There's a very financial sort of people who do something to get rich. He never had that spirit."

Jerry Mander visited Tompkins at Esprit for lunch often, and he found his pal increasingly unsatisfied. "He admitted that he was making a lot of money, and that's good, but he wished he was doing something else," said Mander, who was so trusted that he often received Doug's confessions. "This went on for years and years. A part of him couldn't stand being in Esprit. He kept saying, 'This is not for me. I've got to do something else. I've got to do what *you're* doing.'"

Mander and his business partner, Herb Chao Gunther, had carved out a niche with Public Media Center, San Francisco's leading—and at that time perhaps only—progressive advertising agency. They developed campaigns for David Brower at the Sierra Club, including the wildly popular slogan mocking a proposed dam that would inundate the majestic Glen Canyon in Arizona. In full-page advertisements they ran the banner headline—"Should We Also Flood the Sistine Chapel so Tourists Can Get Nearer the Ceiling?" Public Media offices became an office-away-from-the-office for Tompkins. He came to study what they were doing, to watch Mander and Gunther in action.

Tompkins dug into a wide range of environmental authors, activists, and philosophers. He peppered these cutting-edge figures with faxes, letters, phone calls, and questions. He frequently developed personal relationships with them. He spent hours a day "doing my scholarship," as he liked to call his time spent discovering new authors or reading up on activists who fought for forests. He admired the front row eco-warriors willing to ram and harass Japanese whaling ships and the young activists who lived 200 feet high in a redwood tree for months to ensure that it wasn't chopped down. Tompkins read a wide selection of books about environmental conservation and mocked traditional university education with a passion. Unwilling to be guided or taught, he trusted what he understood, and read.

Tompkins then discovered Deep Ecology, a philosophical call to arms defined by the American writer George Sessions and the Norwegian mountain climber and philosopher Arne Næss. In his work, Næss wrote that humanity required a wholesale rethinking of our place on Earth, and he argued for both a cohesive understanding of ecosystems and an end to human hubris that put one species above all others. The way Næss saw it, the destruction of wildlife and of their habitats was upending the natural balance of life on Earth and sending the entire planet hurtling toward crisis. Decades before climate change and extinction crisis became mainstream topics of conversation, Næss and his band of followers posited that the dream of perpetual economic growth on a finite planet would inevitably—and sooner rather than later—lead to environmental catastrophe. Tompkins was smitten. He fired off letters to Næss and Sessions.

Just five years earlier Tompkins had seduced the deans of design in Milan, and now he was courting the leaders of Deep Ecology. "I was spending my mornings totally immersed in activism, then jolted back to reality by noon and having to concentrate on running the business. Something had to change," said Tompkins. "I set about extricating myself from the business with the aim of dedicating my life to conservation and environmental work. I still wonder how I could have been so distracted by a successful business, so focused elsewhere that I was not out there with the Earth Firsters where my heart actually longed to be."

Part II

Chapter 6

Where Is My North? Flying South

By the mid 1980s I had slowly come to the realization that I was doing the wrong thing. I was in an apparel company making a lot of stuff that nobody needed. My main work was adding to the environmental crisis rather than help revert it. I realized I had to do something else.

—DOUG TOMPKINS

On Earth Day 1989, Rick Klein landed in Los Angeles after an overnight flight from Chile. As soon as he grabbed his bags, he rushed to his first appointment, in Ventura, an hour north, depending on the traffic along the Pacific Coast Highway. He was going to visit Yvon Chouinard at Patagonia, to ask for money to protect a unique grove of forest in southern Chile. "I was carrying around nine different Kodak snapshots of the Araucaria trees that I had taken from the pinnacle of the Cañi Sanctuary. That was my funding pitch. I opened up the nine laminated images, full of beautiful lakes and Araucaria trees, and I said, 'This is what we can actually do now, for $130,000. For that, we can own this and create Chile's first nongovernmental park.'"

Rick showed Yvon his nine photographs. Taped together, they were about two feet long, and they opened like an accordion. The Araucaria trees were as majestic as the redwoods of California. Rick told Yvon it was important to preserve this stand of old-growth forest. He told Yvon he had received one donation so far, of $50,000. Yvon said, "Okay. Here's forty grand. And go tell my friend Doug Tompkins in San Francisco that I just gave $40,000. I guarantee you, he'll try to keep up. That should get you to $130,000."

Klein flew to Northern California, and when he got to town he called Tompkins at Esprit. Klein corresponded through Doug's personal project manager, Dolly Mah. She told Klein, "Doug's interested. If Yvon's in, Doug's in for $40,000. Here's the check." Klein suddenly had money to pursue his dream of saving ancient forests in Chile. He couldn't believe it—"Boom! It was done."

Klein was a native Californian who in the mid-'70s had moved to Chile. He managed to survive the bloodiest years of the military dictatorship by working as a park ranger in Reserva Galletue near the headwaters of the Biobío River. Sporting a shaggy head of hair and a wild beard, he lived in constant danger that the secret police would mistake him for a guerrilla and kidnap then torture him. His boss pleaded away his moustache and beard, saying, "Rick's not a revolutionary, he's a poet."

Splitting his time between Chile and California, Klein ran Ancient Forests International, a nonprofit group based in Arcadia, California. He organized activists to protest against clear-cut logging operations. In Chile, he was fighting to save an endangered forest threatened by Fletcher Challenge, a notorious New Zealand logging outfit. Without immediate action, the Araucaria forest would be clear-cut. But the Araucaria forest in Chile was not the only forest nor the only species of tree that Klein was scrambling to protect. After picking up the $40,000 check from Doug's office, he went to meet with the former head of the Sierra Club, David Brower, who was working at the Earth Island Institute. Klein briefed Brower about his ongoing "Save the Ancient Forests" campaign, which included organizing expeditions that brought activists to the Chilean forests that were facing logging threats along Patagonia's northern border.

Brower had helped on previous expeditions that Klein ran in the

Chilean forests. On one journey they sought to find the "oldest tree in the world," which Klein believed would be an Alerce, a tree nearly as towering as the California redwoods. Klein had drummed up *National Geographic* magazine's interest in the story, but in order to close the deal and to publish the story, the editor demanded a top-notch landscape photographer. Over lunch, Brower introduced Klein to Galen Rowell, one of the magazine's star photographers. Rowell quickly accepted Rick's invitation to join an expedition and photograph the Alerce forest because, as it turned out, in a few months, he was heading to Chile anyway, traveling by small plane, along with his friend Doug Tompkins.

Rowell suggested to Tompkins that they explore the ancient Alerce forest in Chile in search of the oldest tree in the world, and Tompkins jumped at the opportunity. The only way he could survive as CEO of Esprit, he felt, was by taking ever-more-frequent escapes. Kayaking to take a population census of beluga whales and monitoring polar bears in the Arctic—those were the kind of meaningful projects that increasingly stole him away from Esprit. While in Chile on this trip, Doug and Galen planned to raft down the Biobío River and climb Cerro Castillo, a spiked peak in Patagonia.

Rowell's wife, Barbara, would also be flying south to Chile, piloting a Cessna 206 identical to Tompkins's own. They planned a flight path from San Francisco to Tierra del Fuego. Doug had flown this route a half-dozen times, but it never ceased to awaken the wild corners of his mind. To share the piloting, Tompkins brought along his good friend and sometimes business partner Peter Buckley. For moral support in this boy's club, Barbara brought along her brother, Bob Cushman, a ski patrol leader and veteran outdoorsman. Klein agreed to meet them down in Puerto Montt, Chile, after they arrived. He would lead them on the Alerce forest trek.

Buckley had just finished rudimentary pilot training and received a temporary license the week before they left. He had only forty piloting hours in the air, but as soon as they took off Tompkins told him he needed to read and handed over control of the small Cessna. "I got to fly the plane a lot because Doug had a manuscript of a new book, *In the Absence of the Sacred*, by Jerry Mander. And all the way down, he'd read aloud," said Buckley. "The book was supporting a lot of the ideas that Doug had about developing a critique of capitalism and about capitalism as another mega-technology responsible for environmental degradation. That started his new life."

We were feeling dizzy, and flying way higher than we were supposed to be without oxygen, at 14,000 feet. We turn on the oxygen at the very last minute, and suddenly everything is great. Feeling sharp. Aware of all the instruments, and Doug perks up, and then two minutes later there was a rupture in the line, the tank drained. Doug starts getting nauseous, and vomiting in the plane. I'm trying to fly. He said, "I'm fucked up" and puts his head up, and he's trying to really focus, because now it's getting a little bit serious. And he says, "Peter. Go through that notch in those hills, I think there's a . . . I think that's . . ." and sure enough, we go through the gap and I see this big air field. And it's completely by itself. Why is this air strip here in the middle of nowhere? It's not on the chart. Doug says, "Land the plane." So I'm going straight in and I'm coming in for a landing, and I'm flaring the airplane, and Doug puts his head up and says, "Fuck!" And we go, "Boing," bouncing down the runway. And Doug—if it was up to me, we probably would have crashed—but Doug somehow got the plane on the ground. And then he lost it. We pull up, and he falls out of the airplane and is retching on the ground. And then, the military shows up and they take Doug away. And Galen is saying, "Wow, this place is great!" We were essentially under arrest. We were under instructions to fly directly to Lima, and turn ourselves in. We take off and Doug says, "Go down there, and stay low." I go, "Okay." So I'm flying down this valley really low, and the radio comes to life, and the people are shouting. They are completely freaking out for forty-five minutes. Instead of going to Lima [Peru], we head to Chile. We went over the border. And we don't have a flight plan, we're not registered, this is horrible, horrible, horrible. And Doug says, "It's okay. The Chileans hate the Peruvians." The guy who was the head policeman in the town was at a wedding or something. Doug says, "I'm going to go into town and talk to him; it's all going to be cool." Doug goes into town, comes back an hour later, he's a little bit drunk. The policeman is way drunk. He stamps all the papers, he slaps Doug on the back, and we take off.

—PETER BUCKLEY

After surviving a lack of oxygen as they crossed the Andes in Peru and Bolivia, the two planes continued south to Santiago, where Buckley disembarked and headed home on a commercial flight. As they headed further south, Galen flew with Doug, and Barbara piloted with her brother Bob in the copilot's seat. Arriving in Puerto Montt, Tompkins met up with Rick Klein for the first time at a hotel in the center of town. Klein trotted out another accordion-like set of photos, this time of Alerce trees. Tompkins was stunned. The forests reminded him of Norway.

The central government in Santiago had spent a century attempting to colonize the area south of Puerto Montt, yet the rainy wilderness was home to fewer than ten persons in a square mile. Even the 700-mile-long "Pinochet Highway"—dynamited through Patagonia in the mad dictator's obsessive 1980s public works push—failed to breach this dense ecosystem. "Most of it is volcano, or rock and swamps up the valleys," said Klein as he described the steep, rugged terrain. "Because of the scarcity of big game, the indigenous people never went back up in there. Arguably no human has ever set foot in the recesses of some of these ancient forest valleys."

But the forests were under attack. Klein pointed out trucks loaded with native hardwoods that, every few minutes, drove by their hotel toward the docks and added their loads to a pile of wood chips five stories high. Colored like a sand dune and shaped like a volcano, the "dune" consisted of Chile's native forests ground into pieces the size of Doritos. A constant caravan of trucks rumbled into town and unloaded the "chips" that were sold to buyers in Japan. In Tokyo, they became raw material for fine paper.

Driving toward Alerce Andino National Park, Klein was bubbling with excitement. Not only was *National Geographic*'s acclaimed photographer Galen Rowell onboard but so was Doug Tompkins, a maverick with cash and a passion to save forests. The drive was sobering. Leaving Puerto Montt, they saw forest destruction everywhere. Dozens of rustic sawmills were producing five tons of wood chips a day. "They're selling our patrimony for $50 a ton, and the destruction is enormous," declared Adriana Hoffmann, a Chilean botanist. "We're fragmenting and destroying habitats that are unique in the world. And what we're destroying is not a renewable resource. Trees thousands of years old," she said, snapping her fingers, "cannot regenerate like this."

Arriving at the trailhead, Tompkins and his fellow expeditioners hiked into the forest with two Chilean guides and enough food for their four-day excursion. Rowell felt ecstatic. The oldest Alerce trees had sprouted at least 3,000 years earlier, and he knew *National Geographic* readers would love the story of the Alerce. On their second night as they ate around a rustic campfire meal near their shelter, an argument broke out. The guides wanted more money, and they put the squeeze on Klein. Tompkins and Cushman told them to forget it. A deal is a deal. When they awoke in the morning, the guides were gone. And they had taken much of the food.

For breakfast, Cushman distributed power bars, and Klein led a hike to a spectacular stand of Alerce, each tree shooting straight as a ship's mast, some 200 feet high. Rowell set up his photography equipment and Tompkins pulled out a little umbrella and began admonishing those around him not to tread on the tiny forest of sprouting ferns. "It was the most splendid ancient garden we'd ever seen," said Klein, who planned to finish the photo shoot and then camp high up in the valley on a ridge. He was told that on the ridge there were a fresh spring and a trail leading down in a steep, twenty-minute hike back to the gravel road. After the photos and the strenuous hike to the ridge, they were stunned. There was no water. No trail. "Yeah, it was twenty minutes to the lake," Klein cracked, "a twenty-minute freefall over the ledge."

Stuck on an exposed ridge with no water, Tompkins exploded. He couldn't believe that Klein had led them into such a morass. "Instead of having a friendly outing there was fear, blame, hunger, dehydration, and we were bone-racked tired and afraid of rolling off the cliff," said Klein. "Doug said he would never speak to me again. His face had frozen into a grimace."

The group had no choice but to descend toward the lake along a steep, sodden ridge crawling with bloodsucking leeches. For an entire day they slipped and lowered themselves step by step. "It was really, really dense," said Cushman. "We had to ease through the bramble, and climb down the logs or go under them. There was so much stuff fallen over. It was rugged, slow going, and everybody was exhausted by the time we got down to the lake."

Driving home they saw truck after truck piled high with logs or chips. The landscape that had trapped them with fallen trunks and canopy forests was being wiped out. In the larger clear-cuts it looked as if a bomb had exploded. Nothing over two feet high had survived the destruction. Doug was incensed. Breaking his silent treatment against Rick, he shook his hand, looked him in the eye, and asked, "How much did you say this land costs?" He kept pinching himself. "You mean, I can get this for $25 an acre?" He could see how fast the forest was disappearing.

Hoping to rope Tompkins into another one of his pet projects in Chile, Rick guided Doug to fly over Cahuelmo Hot Spring, a backcountry jewel he coveted. After several passes overhead, he directed Doug to land in the pasture of a small farm in a valley just off the ocean. To clear the runway, Tompkins buzzed the pasture to scare away cows and sheep grazing on the landing strip. On his final approach, a group of Chilean cowboys appeared on horseback and chased off the stragglers. As he landed the plane, the five *huasos* galloped alongside the airplane, decked out in ponchos and stiff-brimmed woven hats. Doug decided to buy the massive farm on the spot. "It was kind of a capricious purchase," he later admitted. "I bought it on a whim, thinking at worst it's sea-level native forest that is worthy of some private conservation." It was a bargain basement buy. For the price of a two-bedroom condo in San Francisco, he now owned an entire ecosystem.

Scoping out the dimensions of this environmental wonderland for sale, Doug gazed into the distance and queried one of the *huasos* who knew the area: "Does the land also include that snowcapped volcano?"

"*Sí, Señor Tompkins*," the caretaker answered. "It includes the volcano."

Tompkins returned to San Francisco feeling less tethered than ever to the corporation he had painstakingly built. He began to restructure Esprit's messaging. Susie had created an "Eco Desk" inside Esprit headquarters, and Doug revolutionized the catalog with environmental inserts that included recycling tips plus a marketing campaign urging people to stop buying unnecessary apparel. He and Toscani produced a disturbing image of a chimpanzee wearing an Esprit T-shirt in a grotesque "unmaking" of the Esprit image. Tompkins also directed his staff to adorn Esprit clothing with the message: "Don't buy our clothes, unless you need them." The subversive messaging, Tompkins knew, was likely to create a marketing sensation exactly because it was counterculture. Stunned company bankers and executives confronted Tompkins; one asked in exasperation, "You're saying they shouldn't consume our stuff?!"

Tompkins loved the hullabaloo. Esprit advertising expenditures topped $25 million a year, and as image director, he had authority to shift the campaigns. "My message was really to ask consumers to buy less, which of course is heresy to the concept of free enterprise," he noted. "The response was incredible. Even the advertising industry picked up on it immediately—comments and articles started to appear."

Doug's marital battles with Susie continued to cause friction at the company's highest levels and dragged Esprit deeper into a rut. Finally, the founding couple was demoted and removed from direct managerial roles. "We joke around here that we moved from Doug's MBA program—management by absence, which is fine for an entrepreneurial company—to a program of management by walking around and working," gloated the new CEO, Federico Corrado. Tompkins saw daylight—a chance to escape the confines of corporate America. In secret, he plotted to dump his fifty percent ownership in Esprit and dedicate his energy to fighting for forests.

Susie Tompkins was also eager to divorce her business interests from her ever-more-estranged partner. She described Doug as "a malignant narcissist" propped up by a long line of sycophants. "The way you get people to follow you is by being arrogant, rude, and inappreciative—and not gracious," she said. "That person wants your approval, and so they keep doing things for you. Narcissists are brand-builders because the brand is themselves."

Financial institutions, including banks, also felt wary of Doug Tompkins. "The bankers regarded Doug as sort of a loose cannon, and he certainly was characterized that way by Susie's representative, Isaac Stein," said Peter Buckley. "They were telling the bankers that Doug was just the world's worst executive manager. And Doug said, 'Yeah, that's true, but I'm not going to be the CEO. I'm the artistic director. Peter is going to be the CEO.'"

To launch a new phase of his life at age forty-nine, Tompkins needed cash, free time, and an exit from the corporate world. He made it appear that he was still enamored of the Esprit business in hopes of sparking a bidding war for his 50 percent share. In June 1991, Tompkins first gave signs he was going to lead a flamboyant relaunch of the Esprit brand

and then suddenly sold his share to his former wife, Susie, and a team of Goldman Sachs bankers. "They were just head fakes. He couldn't just sit there and say *I'm going to relaunch the company*. He had to make it *appear* like he was sincere," said Buckley, who was on the Esprit board at the time. "But at no point, no way, no time, did Doug ever intend to buy that company." Doug liquidated his 50 percent share of Esprit's US operations for roughly $150 million. Running the nearly billion-dollar-a year Esprit empire with operations worldwide was no longer his problem.

Even after cashing out of Esprit USA, Tompkins conferred weekly with accountants, lawyers, and tax law specialists. He needed to organize the sale of his minority ownership position in Esprit Europe and also the Esprit Far East divisions. With the Esprit brand powerfully placed in Germany, Italy, and Hong Kong, these shares could allow him to deposit an additional $50 to $150 million, making his total take from Esprit approximately $300 million.

Two decades earlier, Tompkins had sold The North Face, said adios to his family, and driven south with friends on a wild adventure through South America. By the early 1990s, he was as radical and revolutionary as ever. He had no commitments, no shortage of ideas, and a bank account stuffed with hundreds of millions of dollars. As Doug cashed out of Esprit, he was signing myriad piles of documents with his lawyers and with Susie's bankers. "They were in their suits and he was in his usual khakis," recalls his then-girlfriend, Catherine Ingram. "One of the bankers asked, 'So, Doug, what are you going to do with all this money?' And Doug said, 'Try to undo everything you guys are doing.'"

Chapter 7

Earth First!

He was very driven. He was just one of those people. There are only a handful on Earth at any given time, I'd say. He had a fire burning inside of him to do what he thought he had to do. That didn't translate as a kind of "Look at me." He'd be happy to be anonymous in a lot of cases.

—CATHERINE INGRAM, author,
In the Footsteps of Gandhi

Within months of selling his shares in Esprit, Tompkins transferred tens of millions of dollars into the Foundation for Deep Ecology. After The North Face and Esprit this was his most audacious idea, a California nonprofit designed as a vehicle to invest his fortune in environmental causes. Under California law, the maximum initial tax-exempt endowment was $50 million. Tompkins maxed out. After twenty years building Esprit into a global brand, he switched sides. Could he now reverse all the environmental damage he caused while amassing this very fortune? Did he have enough money to make the world a little bit better?

The Foundation for Deep Ecology was a godsend for a select group of environmental activists coming of age in the early 1990s. Seed capital grants ranged from $3,000 to $300,000, as the nonprofit spent millions of dollars every year to fund activist conferences, wildland preservation campaigns, and attempts to slow down what Doug saw as the rampant destruction caused by unchecked global corporations. "Above all we believe that nature comes first, that we are bound ethically to share the planet with other creatures," Tompkins wrote as he sketched his plans for his nonprofit's funding priorities. "We will strive to embody ecocentric, not anthropocentric, values."

Tompkins asked Ernst Beyeler, the Swiss art dealer, to auction the art. Beyeler was stunned when he inventoried his friend's private collection. Not only was Doug's art radar finely tuned—he bought masterpieces by Bacon and Botero decades before they hit mainstream—but he honed an even finer eye for individual pieces. Beyeler raised $18 million selling the art collection, and in a gesture of support for Doug's conservation campaign waived all fees. One hundred percent of the proceeds went into Doug's deep ecology dream fund.

The grants were divided among dozens of cutting-edge groups that Doug and his team found worthy. Or at least Doug. Convinced that loss of biodiversity was "the mother of all crisis," he assumed that environmental restoration was a centuries-long play, a game where final scores would be tallied in the year 2100 or perhaps even 3000. He was equally certain that without a vigorous defense of Mother Earth, there wouldn't be much left to save by 2020, a date many environmentalists held out as the last year to change before critical loss of species and habitat tipped into an exponential extinction cycle. And from his perch—the crow's nest office atop San Francisco's Russian Hill with sweeping views of the Golden Gate Bridge and the waters of the Pacific Ocean—Doug was aghast. The more he read, the more he dreaded the ongoing destruction. Doug loved a quote from Henry David Thoreau, "What's the use of a house if you don't have a livable planet to put it upon?"

Although his mailbox was awash in invitations to ceremonial dinners, awards ceremonies, and art openings, he ignored it all and instead focused on his reading list. Piles of books laced his dining room, kitchen, and bedroom. Even in his plane he often brought a dense treatise on environmental philosophy or stacks of magazine articles with the latest grim news from the extinction front. But rather than let the gloomy predictions dent his enthusiasm, Doug used them to fuel a new passion: he was determined to be the most well-read environmental activist in the room—and he approached the task with the discipline of a man training for a triathlon. With the same passion that infused his love of design, architecture and quilts, Tompkins surrounded himself with the intellectual tools to better comprehend the ecological meltdown that he saw in every country he visited. The guest house on his San Francisco lot became a buzzing hive of activism as Doug assembled a crack environmental team including many young activists.

While interviewing candidates to run his new foundation, Tompkins had fallen in love with one of the applicants. His relationship with author Catherine Ingram was like none other that he had ever experienced. She was a Buddhist, as well as a dharma teacher with an emotional skill set perpendicular to Doug's meager application of empathy. She led him on an inner journey. "We were invited to a very small, private lunch with the Dalai Lama in San Francisco at a museum. It was the first time Doug met the Dalai Lama," said Ingram. "Doug was fascinated by him. He knew he was in the presence of somebody unusual. I brought the whole dharma element into his life, which he loved."

With Ingram, Tompkins had epic battles and a deep transformation. "She's introducing him to a softer side, a more philosophical side, a more spiritual side of the world," said his friend Edgar Boyles. "And it's just at this time when he does not have the obligation to business, and he hasn't quite yet decided the path that he would eventually take. The people that she introduced him to had something to do with the epiphany. I know there was some ayahuasca involved, as well. She was connected to a different area of the world, which was somewhat astonishing to the Doug watchers that we were. *Wow, this is different, she's different.*"

Ingram preferred comfortable hotels; Tompkins a sleeping bag on a random couch. He was impulsive and stubborn. She was the author of

In the Footsteps of Gandhi and books on dharma. They thrived in the moment, but he could never settle down. "He couldn't quite do it; it was a bridge too far," she recalled. "A couple of times, I led him in guided meditations; did the Big Mind guided meditation now and again. He would say afterwards, 'Yeah, I should do more of that.' But somehow life was always galloping along with him."

With Ingram motivating the ride, Tompkins launched an impromptu whirlwind world tour, one month chasing Deep Ecology founder Arne Næss across Norway, the next traveling to Mexico with the Anglo-French environmental philosopher Edward "Jimmy" Goldsmith. "Jimmy invited us to his pad in Mexico, this extraordinary place," said Ingram. "And we brought Jeremy Rifkin, Norman Lear, Richard Branson, Jerry Mander, and others for three days to strategize a plan to start challenging globalization."

When he wasn't with Ingram, Tompkins pined for her. Tompkins shocked his longtime aide Tom Moncho by narrating their road trip across Norway with sentimental love stories and a Van Morrison ballad that reminded him of her and that he played over and over on the eight-track. "We actually fell kind of crazy in love, I have to say," said Ingram. "It was one of those things. You know, there is a Shakespeare line, 'Whoever loved that did not love at first sight?' That was kind of how it was for us."

As his environmental grantmaking took shape, Tompkins turned over day-to-day operations of his foundation to his eldest daughter, Quincey. "When I was a teenager, Esprit was the thing," she said. "I was the first in my family to go to college. Doug said, 'Just go; just go study language, go study art, go study culture, history, or whatever, and then come run the company.' Yeah. Sounds good to me. So I worked at Esprit, on the environmental desk. But then a couple years later, he said, 'Come and run the foundation.' Well, an environmental foundation? This was new."

Convinced that mainstream environmental groups were as complacent as they were complicit, Doug searched out those activists who threw their bodies on the line, the kind of protesters who might chain themselves to a bulldozer—or, like Doug, Yvon, and other guests at a memorable Thanksgiving party on the beach, pull up acres of survey stakes to slow down construction of a new highway.

Pranks and media hijacking pleased Doug. Always a fan of circus acts, he appreciated the humor when activists unrolled a "crack" three hundred feet down the face of the hydroelectric dam in Glen Canyon, Arizona. Of all the dams built in the western United States, few provoked such outrage as the $800 million cement structure that drowned Glen Canyon and created artificial Lake Powell, 186 miles long. From a distance the unrolled plastic looked like a crack, as if the structure was buckling. The "crack" was actually a strip of dark plastic weighted to roll hundreds of feet down, like a burgeoning break.

These kind of pranks and public relations stunts were immortalized in Edward Abbey's 1975 novel *The Monkey Wrench Gang*, in which a band of merry pranksters wage war on industrial machinery under their slogan "No Compromise in Defense of Mother Earth." The hero is George Hayduke, a Green Beret Vietnam veteran who rebels and takes the side of nature. The "comic extravaganza" novel became a sensation, with a following that inspired a real-life movement dedicated to catapulting Abbey's "monkey wrenching" from fiction to fantasy to fact. They called themselves Earth First and their logo always carried the exclamation point, as in "Earth First!" These were the little-known but highly effective corners of the environmental uprising that Doug Tompkins gleefully seeded. "Doug helped this magazine called *Ad Busters*," explained close friend Edgar Boyles. "He funded the magazine; it was based out of Canada. It was what they called a 'culture jamming' operation, where they'd do fake ads and posters that challenged people's norms and perceptions about consumption and overconsumption."

Captain Paul Watson was another activist who found in Doug Tompkins a passionate ally. Watson participated in the founding of Greenpeace with a half-dozen fellow activists, then broke away. In 1977 Watson assembled Neptune's Navy, a motley flotilla that would eventually include a catamaran and a pair of run-down icebreakers, useful as battering rams. With his rebel armada, later renamed the Sea Shepherds, Watson launched a campaign of resistance against whaling ships. His declaration to disable "the enemy fleet" inspired followers, donations, and worldwide sympathy that was later enshrined by *Whale Wars*, the Animal Planet miniseries. "The Sea Shepherd crew is doing what governments should be doing, but refuse to do themselves, because of the threats of trade retaliation from Japan," said Watson, who for years received donations from Doug Tompkins via the Foundation for Deep Ecology.

"Doug was never afraid to fund things. If anything, it was kind of my job, together with the CFO and the lawyers, to make sure this is okay, like legit," laughed Quincey Tompkins, his eldest daughter and executive director of the Foundation for Deep Ecology from 1991 to 2000. "There was nothing illegal, just things that were radical. He would support those guys. And we'd have to vet it."

The first act of tyranny is when large corporate entities in a society try to privatize the public trust and steal it. In a true free market, you have to properly value natural resources, and it's the undervaluation of those resources that causes us to use them wastefully. What polluters do is they make themselves rich by making everybody else poor. They raise the standard of living for themselves by lowering the quality of life for everybody else, and they do that by escaping the discipline of the free market and forcing the public to pay their production costs. If you show me a polluter, I'll show you a subsidy. I'll show you a fat cat using political clout.

—ROBERT KENNEDY JR.

As Tompkins bankrolled first dozens and then hundreds of direct action conservation groups via his foundation, a campaign against environmental activists erupted in the western United States. In Moab, Utah; Ukiah, California; and Missoula, Montana, environmental activists were threatened, shot at, had their dogs poisoned, and their homes torched. Funded by the coal industry, big oil, and snowmobile manufacturers, an aggressive antienvironmental campaign called "Wise Use" erupted as a countermovement to the rising environmental tide of the early 1990s. Posing as a grassroots organization, "Wise Use" was later exposed to be largely a PR effort funded by polluting industries.

At the same time, Dave Foreman, one of the founders of Earth First!, was targeted by the FBI in Operation THERMCON. After a year-long sting effort and more than a thousand hours of phone intercepts, along with attempts to lure Foreman into illegal acts, FBI agents arrested him as a SWAT team surrounded his modest New Mexico home. "I never felt so naked," quipped Foreman as he described awaking, nude, with his wife, Nancy, to find agents pointing a pistol at his head.

The FBI and federal prosecutors tried to frame Foreman by alleging, on scant evidence, that he was part of a conspiracy to sabotage electrical power lines leading to nuclear power plants. Federal prosecutors sought a twenty-year prison sentence for Foreman. The sham trial might have prevailed had Doug and Yvon not helped fund Foreman's defense. As the accused activist fought the bogus charges, he and Tompkins spent hours discussing what they both saw as the FBI's war of "dirty tricks" on Earth First! Tompkins and Chouinard paid tens of thousands in hotel bills and bought plane tickets for the legal team, led by Yvon's neighbor, the skilled defense attorney Gerry Spence.

The defense strategy paid off. After months of procedural delays, the FBI was forced to hand over clandestine audios of talks with Foreman. During the final push to frame him, the FBI agent Michael Fain chatted with a fellow agent during the operation and forgot he was recording. "This [Foreman] really isn't the guy we need to pop, I mean, in terms of actual perpetrator," said Agent Fain. "But this is the guy we need to pop to send the message and that's all we're really doing and if we don't nail this guy [Foreman] we're not sending any message. . . . Uh, oh! We don't need that on tape. Oh, boy."

Foreman was declared not guilty of all serious charges. He then celebrated with a nationwide collegiate speaking tour during which he warned Green activists to prepare for "serious prison time" and asked if they were willing to "put your life on the line." Foreman was an inspiring orator. Dressed in blue jeans and cowboy hat, he encouraged activists to pull up survey stakes, criticize mainstream environmental groups, and defend Mother Nature at any cost. "The great thing about Doug and Yvon," said Foreman, "is the consciousness that making money isn't an end, that you need to make money so you can accomplish something with it. Doug was very much focused on the small groups, the hard-hitting groups."

By the early 1990s Earth First! was in crisis mode. FBI provocations (including undercover agents having sex with activists, buying explosives, and provoking violence) sent the paranoia factor through the roof. "Earth First! was founded to save big, wild places, and take no-compromising stances on protecting wild nature," explained Foreman. "It got infiltrated by the FBI, because apparently the Reagan administration saw us as a real threat. Earth First! was also attracting people who were countercultural, against the state—against the man, so to speak—but who didn't necessarily have a strong wilderness ethic, so the old guard broke away from Earth First! and started a magazine called *Wild Earth*. Right away Doug Tompkins sent a nice letter and a big check."

To expand his conservation plans, Doug turned to two voices he respected and trusted in the environmental movement: those of Foreman and John Davis, both central cogs in the Earth First! organization and the *Wild Earth* brain trust. "Doug asked me, 'Who are the best activists?'" said Davis. "I don't think he told me much about himself.

He just told me that he was a successful businessman and interested in investing in effective wilderness and wildlife groups, and who did I think was the best? I spent some time with him on the phone talking to him about which groups were very effective at defending wild places and wildlife."

From the office inside Tompkins's San Francisco property, Davis, Foreman, Jerry Mander, Quincey, and Doug honed their environmental agenda. After years at Esprit, dreaming of an escape from corporate life, Doug was a "free radical" with $150 million in cash. Wasn't anything possible?

Tompkins challenged Foreman. "You've been talking about North American wilderness recovery, you've been talking these big ideas for years, they're great ideas," he said. "Why don't you implement them? Why don't you get the key people together, make a plan to implement this North American wilderness recovery strategy?" Foreman took the bait and began preparations for a forest restoration summit. Tompkins paid airfare and food. Lodging was on his living room floor.

At his hilltop home in San Francisco, Tompkins gathered scientists, authors, young forestry activists, grizzled vets like Foreman, and top biologists, including Michael Soulé, the "intellectual grandfather" of the scientific field of conservation biology.

Evening after evening, the living room overflowed with passionate debate. Tompkins loved to mix up the crowd, adding friends like the Italian jean designer Fiorucci and Randy Hayes, the founder of Rainforest Action Network. Although he could harangue a table of VIPs with no pause for diplomatic niceties, when meeting the activists he could also listen. "Doug knew that there were people around him that knew more, read more, and were more experienced in these nonbusiness issues than he was," said Hayes. "With all of his energy, and his curmudgeonly way, he was actually very humble in his relationship to us, to those around him he considered to be people of knowledge and thought that he could learn from."

"If you were in a situation where you need to achieve something by compromise, that just wasn't the outcome that interested him at all," said Rick Ridgeway. "He didn't have any interest in trying to bring unaligned people into an alignment. But he's good at bringing like-minded people together to reach a common goal."

Doug's overdose of ego, along with his cocky and blunt way of talking to people, sometimes meant he was described as an egomaniac. But he often stunned his friends and enemies alike with his ability to sniff out cultural trends long before they were recognized. As a global businessman, his nose for the future provided him with hundreds of millions in wealth. Now, having converted his house into a hot spot for environmental thought and activism, he asked himself big questions: What are the boldest plans to save intact forests? What are the most destructive clear-cut practices? With tens of millions of dollars, how could he slow industrial destruction of ecosystems?

When a forestry activist trying to shame British Columbia into action needed a place to crash, Tompkins allowed him to sleep on the couch at Lombard Street for weeks as he strategized how best to rebrand his province. When together they pushed the slogan "Canada, the Brazil of the North," Doug was delighted. He loved to attack corporations with clever phrases. Especially if they contained humor. "It was a comfortable setting. People felt empowered. You had a bunch of people just like Doug, who wanted to make something happen," remembers Andy Kimbrell, a sustainable agriculture expert who attended activist summits with Tompkins at his home.

With vanguard conservation ideas top of mind, Tompkins funded a range of activities that spanned from three-day strategy meetings to lawsuits that might last years. He considered litigating for wilderness and wildlife to be one of his most efficient investments, and he routinely poured hundreds of thousands of dollars into legal battles around the globe. Although some legal protection for forests, rivers, and wetlands had existed for decades, environmental activists were routinely outspent and outgunned by corporations with white-shoe law firms vying to snare their accounts. Few grassroots groups could endure a years-long legal siege that cost a million dollars. How could a neighborhood environmental defense group—on a garage-sale budget—triumph over pockets that deep?

"Doug put money down on big ideas. And he was challenging the system with a lot of it," said Victor Menotti, an author and antiglobalization activist who brainstormed with Tompkins at his home on Lombard Street. "Doug was unique in supporting thought leaders, bold-ideas people, giving them the space to come up with plans, articulate agendas that were impactful and unique and so badly needed. You're taking a risk on an idea, and you hope that they pay off. It was kind of like venture capital."

While ignoring five-star hotels and flying economy, Tompkins allowed himself some luxuries. He rarely went a couple of days without jumping into the cockpit of one of his airplanes. Sometimes it was for business, but usually it was to visit a property he was considering buying. On a trip to British Columbia with environmental author Doug Peacock, Tompkins overflew intact forests, relishing mile after mile of untouched trees. The forests were so thick that from the air Tompkins could rarely see the ground. Then he saw the scar. Running parallel to the majestic trees was a tangle of mangled stumps, shattered branches, and tire tracks. "We were both awestruck by the extensive destruction of native forests we saw below," he lamented. "We watched with sadness and despair the landscapes below laid waste by industrial and technological managers who had clear-cut vast tracts of ancient forests. As I gazed out the windows of our small plane, we withdrew into ourselves, reflecting on how one person could help stop this madness."

A vision formed in his mind: Tompkins would design, then publish, a book on forest conservation. As a pilot, he saw how forestlands were cleared away. It looked as if a giant had taken a razor and shaved the green fuzz from the skin of the Earth, leaving only stubble.

He was determined to divorce his actions from anything close to mainstream sensibilities. He saw many environmental NGOs as soft, ineffective, and unable to truly appreciate the wild.

> *Those people are all city oriented, they don't have much. . . . They look around and the landscapes are just scenery. They don't see it as an integral whole. They're believers in technology, so they believe in industrial forestry, industrial agriculture, industrial fishing, and all this stuff, because they are technology optimists. They really think this stuff is all okay; we've just got to reform this a little bit, or this other. But they don't have a deep, systemic analysis of the crisis that we're all ensnared in. So, they put their efforts, as I see it, in the wrong places. You could call it shallow ecology. They're reformists. Human welfare environmentalism: which sees nature as a vast storehouse of resources for human use. They don't have a deep respect for other creatures. They don't really see that sharing the planet with other species is fundamental: a point of departure in order to orient ourselves toward our human economies and social structures.*
>
> —DOUG TOMPKINS

Twenty-five years earlier, Tompkins had studied the activist books published by the Sierra Club, which were the brainchild of David Brower, whom John McPhee had labeled the Archdruid of environmental activism. "One of his great legacies was the Sierra Club book program, from which I learned a lot and which was the inspiration for our own books," said Tompkins. "I owe Brower all the credit for sowing that idea in my head. I thought that if the Sierra Club could do that for beauty, we can use that same process with ugliness."

The Foundation for Deep Ecology granted a quarter million dollars to launch the forestry preservation project with a coffee table book that thunked down at eight pounds and was entitled *Clearcut: The Tragedy of Industrial Forestry*. The book was a beautiful presentation of all that is ugly about clearcutting old growth forests. Hillsides, valleys, and checkerboard cookie-cutter scars the size of football stadiums. Doug never planned to sell the book; he would give it away. It was meant as a gift and a tool in the larger public relations campaign to denounce destructive forestry practices. *Clearcut* would illustrate how timber companies destroy mile after mile of native forests, then leave "beauty strips" along highways to deceive the public about the extent of forestry destruction.

Tompkins was a pilot and knew these truths; in the book he wanted to bring the message to "flatlanders" traveling in autos who rarely knew that the gorgeous trees lining the roads were not the entryway to a forested paradise but a curtain hiding massive destruction. The book design was placed in the hands of Tamotsu Yagi, the award-winning designer who spent years working with Doug at Esprit and later for Steve Jobs at Apple.

When the mock-up for *Clearcut* was finally ready, Doug asked for each page to be blown up to a 16 × 20-inch image—he wanted to walk his mind through the pages. Together with Yagi, Tompkins gathered at the Japanese designer's San Francisco studio. As Tamotsu and Doug rearranged the layout of the photos and eagerly tweaked the overall design, a seventy-two-year-old gray-haired and slightly stooped mountain climber showed up. It was David Brower, the veteran of World War II and then countless battles to create national parks in the US. Brower had been booted from the presidency of the Sierra Club and went on to found Friends of the Earth and Earth Island Institute.

Brower jumped into the critique. The three men paced the room as they discussed each photo. "Doug was forty-nine years old. And Brower was in his mid-seventies, white haired. I sort of stood back, and I could see the wheels turning in both their heads," said Edgar Boyles, the photo editor for *Clearcut*. "Brower made books about beauty to try to save the world," said Boyles who instantly realized what his friend Doug was doing. "Here was this book which was going to be this beautiful book, about destruction. I saw a passing of the torch."

When *Clearcut* was ready for printing, a controversy erupted. North American printers refused to print the tome. They wanted no part in the project. Trashing the pulp and paper industry was not in their interest. They were afraid of a backlash. Exposing companies that chopped down old-growth forests and planted tree farms to make paper would be commercial suicide.

Finally Doug and Tamotsu found a printer in Japan willing to publish the book. The result was stunning. Each copy of the book arrived in a custom-made box. The cover was straw board while interior pages were printed from postconsumer paper. Essays by Bill Devall, Michael Soulé, and other leading thinkers on the future of wildlands were interspersed with apocalyptic images of destroyed forests. Printed in both hard- and softcover editions, thousands of copies were given away, including to every member of the US Congress. Public libraries across the nation and activists working on the frontline of wildlands conservation also received a copy. "It was under the mattress of students studying at forestry colleges. It was almost a banned book," laughed Boyles. "And of course, it was checked out of libraries in Montana, never to be returned. It was checked out and destroyed."

To many environmental activists, including Doug's close friends, the production costs for the book were exorbitant. Doug, however, measured the worthiness of his investments on a different axis. His "balance sheet" valued the recuperation of apex predators or the projected health of a forest some 200 years later, in the year 2192. A redwood tree that sprouted in 1700, for example, would not reach middle age until roughly the year 3000. By the end of the redwood's life it would be closer to the year 4000. That was the kind of radical thinking that propelled Doug Tompkins out of bed every morning. "I always did more than anybody else I knew, with few exceptions," Doug told an interviewer as he approached his fiftieth birthday. "I don't have the sense of not having done something, or that I should accomplish more in my life, or know more experiences—because I do them all the time. That's what staves off any of this classic midlife crisis syndrome. I've already led the lifetime of ten people."

Doug actively refused to get old. His energy level was brutal. Before his friends were awake, he'd bike from his hilltop San Francisco home, across the Golden Gate Bridge, into the Marin Headlands, and after bouncing along ridgelines would then complete the full roundtrip and be home before 8:00 a.m. "Doug, in a day, he would do what I would do in a week or two," said Ingram, who lived with him at the time. The energy was both wildly attractive and fiercely independent.

Surrounding himself with a cadre of young thinkers and activists propelled Doug forward. He was as far from the country club set as humanely possible. "I go to the Telluride Film Festival but I sleep on the couch of my friends that are running the festival," he said while confessing that he thrived on the buzzing energy of youth. "So many kids—far more than the average adult—have bright and sparkling eyes. They're really receiving stimuli; they're taking things in from all over. You can see it. It's all in their eyes. They're looking everywhere. Stuff is coming in unfiltered . . . and then you'll see some parents that have got these kids dulled down so fast, things are coming in *very slow*."

Tom Butler, the longtime editor of *Wild Earth* magazine, spent decades with Doug working on books and publications as his editor-in-chief. Butler, who helped Tompkins publish more than a dozen different coffee table books (each as large and exquisite as *Clearcut*) described the strategic thinking behind the Tompkins publishing empire. "Doug thought that the scrappy grassroots wing of the conservation movement, because it was undercapitalized, under resourced, was never going to have the money or the creative capacity to produce an oversized coffee table book on the damage caused by mechanized forestry on public land," said Butler. "But if they were effectively given that tool, they would have, in his phrase, 'a paper tiger.' You look different if you go into a meeting as an activist with a book that clearly has high production values. Your political opponents take you more seriously than if you're just some scruffy dirtbag conservationist."

Over the course of their joint publishing efforts, Tompkins and Butler produced a library shelf full of books ranging from criticism of industrial animal farms to suggestions about how to control wildfires. The final bill for the book publishing was just over $13 million, and Doug never regretted for a moment his bet on books.

"Doug was like an Indian rishi," said Vandana Shiva, the food-sovereignty activist who participated in the San Francisco strategy meetings at Doug's house. "Rishis are the sages who recognize that the world in which we live, the material world, is a passing one, and they dedicate themselves to the lasting. They do it by having very little of their own. He shifted his money to live the ultimate life of giving. Giving to nature. To protect nature in a time when nature is being trashed. For me, shifting money from growing profits to growing nature is the highest level of human elevation, of consciousness."

As his conservation plans took form, Tompkins prepared to migrate to Patagonia. As he prepared to leave San Francisco and move to Chile, Tompkins held a yard sale and sold his Lycra cameras, Amish furniture, and Lange ski boots, vintage from the days of The North Face. He prepared to abandon his chic San Francisco home and cede his ranking on the San Francisco social scene as he relaunched his life in a direction that even he couldn't begin to explain. His Lombard Street house would continue to host the foundation, but Doug was migrating. He took his girlfriend Catherine Ingram to southern Chile, flew her over the farm he'd bought, and suggested to her it could be their new home. "I could see he was having a love affair with South America," said Ingram. "Everywhere we went, he was raving about, 'Taste these tomatoes.' Everything was so beautiful for him about Chile." Ingram refused to join the mission. They disagreed on hemisphere. She didn't speak Spanish and thought British Columbia sufficiently remote.

She appreciated that Tompkins called her "the best indoor girl" he ever dated. But when frustrated, Tompkins would say that of all the women he had dated, Ingram was "the worst outdoor girl." They broke up just as Tompkins finalized his plans to escape to a remote corner of southern Chile.

"Doug used to say to me, 'I expect to die in an accident someday,'" said Ingram. "And we would have these terrible conversations in which I would try to convince him that his life was more than just his own, that many others cared for him, and that my life would be forever affected in that loss. You don't casually play around with life! And he'd say, 'My life isn't worth living if I can't do the things I love.' I would argue, 'Well, aren't there ordinary things that you love well enough that wouldn't require risking life itself?' But I could never prevail in this—not for one minute. He would never concede that this was an option."

Chapter 8

Pioneer Village

There wasn't a master plan, but this wasn't a guy who was throwing darts at a board. We had many conferences at Lombard Street where we'd bring in the top environmental thinkers, the top conservation biologists. He was informing himself. He was learning. He was thinking about big, connected, wild, beautiful areas. He was nearly fifty years old. He was ready. He was shifting from a business that didn't fit his world view anymore, and yet, he couldn't ever get the empire building out of his system. He was an empire builder. He couldn't stop. Ever. That was his deal.

—DAN IMHOFF, Esprit environmental desk officer and son-in-law of Doug Tompkins

In 1991, Tompkins flew thousands of miles to Santiago, then another two hours south to Puerto Montt, where he piloted a small plane to the run-down Reñihue farm with the snowcapped volcano—this was now his home. None of the buildings in Reñihue were near-waterproof or capable of holding heat. While surveying his ragged farm, Tompkins found an eight-foot-by-eight-foot structure. Itinerant shepherds found it useful to survive the rain. Mounted on wooden skids, this tiny house had once been dragged by oxen through fields hacked from thick forests. Without electricity, candles and oil lamps provided light. Refrigeration came in the form of a metal box strapped to the outside of the hut and cooled by winds from nearby ice fields. Drinking water was collected on the roof and dripped into a wooden barrel. Instead of the Golden Gate Bridge, Tompkins now had a view of Michinmahuida, a volcano, which was blanketed in snow and prowled by pumas.

Doug sought out local artisans and renovated the tiny shack, added a second tiny house, and installed a wood stove to replace the fire pit, then created a freshwater collection system that funnelled water into the kitchen and an exterior bathroom. Even Doug had to stoop to pass through the front door. Friends called the shack the "Hobbit House."

A caretaker led Tompkins by horseback on a tour that showcased the rivers, lakes, and miles of coastline he now owned. There were few settlements larger than fifty inhabitants for a hundred miles in any direction. Ferns with six-foot leaves sprouted from the mossy forest floor. With as much as two feet of rain *a month*, the region was shimmering green, with shrubbery so thick that no ordinary-sized deer could penetrate the thickets. Just thirteen inches high, the endangered *pudu*, the world's smallest deer, was endemic to these forests.

Tompkins loved the pioneer experience. By day, he kayaked the remote fjords swarming with black-and-white dolphins. He ice-climbed to remote glaciers with friends when they could penetrate the weather to visit him from the Northern Hemisphere. When hungry, he fished from either riverbank or kayak, and he harvested vegetables from his recently constructed greenhouse. The rain and cold were so persistent that a wood stove was required to keep the plants alive even in the greenhouse. When ocean storms reached a peak, Doug launched a wooden-hulled boat into the raging waves of the Pacific.

Decades of generous government incentives—including free title to land, subsidies for home heating, and exemptions from sales tax—did little to boost population. Along the route, abandoned human settlements lay scattered like seeds that never sprouted. Few towns registered more than a thousand inhabitants. Several areas, like La Junta and Cochrane, grew, but they were few and far between. Summertime visitors to the region were measured by the dozen. In winter, mud and landslides made the road impassable for weeks. Hundreds of rivers descend the rugged terrain, most mapped N/N—"No Name." Given the Chilean's love of placing numbers on the map, each slice of Chile was given a Roman numeral. Reñihue was part of the tenth region and designated simply "Region X."

Ferryboats passing by the area offered transport north to Puerto Montt, yet few frontier families could afford these expensive journeys. Unless they needed to visit family, buy provisions at a proper store, or seek medical attention at a decent hospital, locals tended to travel infrequently, rarely straying more than 200 miles from their humble outposts. Sunday horse racing was an excuse for gathering at the impromptu racetracks cut from the fields.

Most supplies that arrived by boat were loaded in Puerto Montt and then after a day of navigation unloaded at a small pier near Tompkins's home on the fjord. This flow of goods and deliveries continued nonstop. But when tides were at their lowest, the boats would beach a mile off shore and Doug, in boots, would wade into the waters carrying crates of supplies on his shoulders.

Despite the outward appearance of neglect and abandonment, Doug was certain this neighborhood held vast potential. Plots of land extended as large as tens of thousands of acres, with the price per acre as little as $25. A backyard the size of Manhattan's Central Park could be bought for $60,000. "Some dude lost $35,000 at the gambling table and was willing to sell Doug a whole valley full of old-growth Alerce trees," said Dan Imhoff.

Tompkins felt enamored of his new home, but he needed to organize logistics with the outside world. So he set up offices and a home in the city of Puerto Montt, a grubby outpost of loggers, fishermen, and ranchers. With packs of mangy street dogs both endearing and ghastly sick, Puerto Montt left a nasty first impression. Aesthetics seemed to have bypassed the mayor's office, the parks commissioners, and the public works department.

The lack of city sewage, proper drainage, and municipal zoning regulations, and salaries that rarely topped $200 a month meant that family life in Puerto Montt carried few signs of the modern era. Even by the mid-1990s color TV had only just arrived. Telephones were still a privilege. Clearly, the city was born from the wealth of the forests, in particular the Alerce wood. Shingles told the story. Overleaved in layers to seal out windblown rain, the shingles came in dozens of designs. A single Alerce shingle could absorb eighty years of rain and still remain solid. Several hundred industrious families from the ruling class built Victorian-style homes in the region. Elaborate doors were carved out of Alerce for the Catholic churches. The distinctive look became known as Estilo Chiloté, in homage to the nearby island where boatbuilders and carpenters crafted hulls and homes by hand. To Doug's deep delight, a love of craftmanship and woodworking was alive. But few residents could afford the labor to cut, carve, and set the Alerce shingles. Most families in Puerto Montt huddled under tin roofs onto which raindrops the size of gumdrops clattered noisily while rivulets of water leaked into their humble homes.

Chile in the early 1990s was a country traumatized. Much of the populace suffered from what might best be described as a collective case of PTSD, following seventeen years of harsh military rule, state-sponsored assassinations, military propaganda flooding the airwaves, secret police hit-squads, and widespread use of torture and "disappearing" one's enemies. The populace had been cowed into if not submission then at least a broad-based conformity.

From 1973 to 1987, the United States had supported the brutal tactics of the Pinochet regime, but in 1986, they smelled change in the air when a team of Cuban government–trained assassins ambushed the general, killing five bodyguards and wounding but not killing the dictator. Clearly his days were numbered. In 1988, the United States finally switched sides and supported the civilian movement calling for his ouster. Chilean political and business leaders were opening to the world, eager to demonstrate that the nation's free-market ideology and "everything is a commodity" mentality would remain immune to the passionate clamor for social change at a deeper, structural level. It had taken nearly two decades of brutal implementation, but the mission was complete—Chile was regularly cited as the "freest market in the world." Destruction of the local environment began to ratchet up.

Despite the deep misgivings about complicity by the armed forces in human rights crime, many residents of Region X felt abandoned by the central government and were thus loyal to the few signs of authority—including the navy and army—who provided rescue services and delivered emergency supplies.

In Puerto Montt, Doug bought an architectural masterpiece known as "Buin House" and set up shop with Marci Rudolph. At Esprit, she had been the showroom guru. Before a showroom would open to the public, it was Marci who came to set the stage. When Tompkins first began talking about jumping ship and launching a new life in remote Chile, she was tempted to join him. And when she got a letter from him, she moved to Patagonia for months at a time, helping Doug.

"We lived with this family and their dogs, their grandmother, and their kid in this wild house," said Marci. "We lived in this upstairs space that I don't even know if it was really supposed to be part of the house. There were three teeny rooms. The oldest daughter, she was out of architecture school, but she lived at home. Then me and Doug. We would share a bathroom. He just thought it was so great. And I was thinking to myself, *What's so great about this?* He loved the weirdness of it all. The grandmother would shuffle to breakfast with the dog, who sat on her lap."

While the Buin House in Puerto Montt was being rebuilt, Tompkins studied local geography, ordering government surveys, tourist maps, highway plans, and geological overlays. He became in effect an amateur cartographer, a student of infrastructure, and a fierce critic of Chile's resource-extraction-based economy. Doug continued to read a book every week, sometimes faster. He devoured scientific reports, and studied development plans for the region. He acted with the passion of the recently converted, yet even his closest friends couldn't figure out *what was his plan?* Was he going to stay in Chile? Would he become bored and move back to California?

Tompkins brainstormed various strategies to protect the forests of southern Chile. The deforestation was running full speed, and he wondered if he might slow the clear-cutting of ancient forests by impeding the expansion of roads into Patagonia. He was convinced that "come road, come destruction." He'd seen it happen in California's Sierra Nevada mountain range, in British Columbia, and in the Rocky Mountains. As he projected the natural extension of roads that eventually needed to cross a river, he calculated the most economical spots to build a bridge. Could he sabotage bridge building? What if he bought up those potential river crossings? Could he slow down the bridge building? If he bought up enough narrow channels, could he stave off industrial forces long enough for public consciousness to catch up with the insanity of short-term profits? It was only a matter of time, Doug posited, until citizen's awareness and outrage aligned with his dire prognosis: that globalized capitalism was killing the planet. His diagnosis pointed to "development plans" and road building schemes as preordained wounds that would spider-web out in all directions, like an infectious disease or growing tumor. Tompkins believed that David Brower was spot on when he suggested that the US government fund a Peace Corps–style program to restore abused ecosystems under the moniker "Global CPR: Conservation, Preservation, and Restoration."

"I had no idea who Doug was," remembered Eduardo Rojas, the award-winning architect then living on Chiloé island. Rojas recalled an out-of-the-blue phone call from "a gringo from California" who in heavily accented Spanish asked for a meeting. "He told me that he had bought Reñihue, and that he was looking for an architectural studio to help him with this adventure. He had been researching, and found a little book I had helped to author. Doug read my name in the acknowledgments and decided I was the guy to work with. He flew in, we met, and later that day we flew to Reñihue. Doug was a person who made decisions quickly and demanded action."

Landing in Reñihue, Doug outlined to Rojas the extent of the neglect. The barn was half collapsed, the cows had escaped to the hills, and the only terrestrial access was a treacherous boat ride, through channels from Puerto Montt. Tompkins loved the challenges of this mission. Like the tiny North Face boutique he had renovated in San Francisco as a twenty-one-year-old, here he had a challenge worthy of his prodigious energy. Could he restore this abused land?

Before Tompkins's arrival, the caretakers and their families living on the Reñihue property had rarely met a non-Chilean. They studied the new owner closely. The first week, he organized a micro-trash pickup. Locals typically buried their trash or burned it in a hole. Tompkins was incensed. He couldn't believe that there were cigarette butts and little pieces of trash everywhere. For the first couple of days he demanded that the staff pick up all the micro-trash. "The people were like: *This can't be for real!*" laughed Marci. "They were cowboys, gauchos—tough guys. They didn't pick up trash."

While he was fixing up his airstrip, Tompkins bought a Husky, a high-horsepower, lightweight airplane able to fly as slow as forty miles per hour, swerve like a bird, and take off or land on anything longer than a pair of tennis courts end to end. Tompkins explored his lands, buzzing along just above the treetops. With extra fuel tanks added to the plane he could fly longer, explore deeper. He entered tight canyons and gaped at the singular beauty of his new neighborhood: active volcanoes, soaring condors, forests never logged, rivers never dammed—all undisturbed, exquisitely designed by nature, without a single flaw. Except for the cows.

"One of the things that drove Doug crazy was these feral cows," said Andy Kimbrell, the writer and environmental activist. "This whole area had been chomped by cattle, to the point of ruining the soil. So, he had a big remediation job. All these cattle guys were gone, but the cattle had gone feral; they'd gone wild. It's sort of like a canyon, a valley between mountains. He would go up in the airplane, I'd go up with him, and he'd fly way in the back where these feral cows were, just hanging out, getting fat and all-ferally. He would chase them flying low in his airplane. If you were afraid of airplanes, you didn't want to do this with him, but I liked it. He would chase them, because they hated the noise. He moved them repeatedly to the front where they could be captured. He loved getting rid of every one of those cows." After corralling the cows, Tompkins packed them on a barge and shipped them back to the mainland where they would be sold.

Bushwhacking this rain forest was nearly impossible, as stalks of *quila*, a native bamboo that could reach seventy-five feet tall, formed walls so thick that they were impenetrable. Exploring by airplane or in his wooden crab boat was faster, but Tompkins enjoyed the silence of paddling. When Yvon visited, they kayaked and climbed for days, camping along the fjord and exploring upriver to a sheltered valley that protected them from the dangers of ocean waves and gale-force winds. Doug sought nature's soundtrack for his deep thinking, and Yvon didn't need small talk to feel comfortable. They'd sometimes go half the day without a word and felt their relationship was ideal.

Just as the digital revolution of the early 1990s exploded, Doug sprinted down the gangplank, against traffic, and went off the grid in an embrace of all things analog. He migrated from running a billion-dollar-a-year enterprise to subscribing to *Lehman's*, the magazine that offered nineteenth-century pioneer products at twentieth-century prices. Where else could he order a kerosene-powered chicken egg incubator? Or choose among a dozen hand-powered apple peelers? "He had these romantic visions, but people wanted progress," recalled Marci. "The caretakers living at Reñihue thought, *Yeah! An American guy is buying the place, we're going to get washing machines!* But Doug wanted to go backwards. And they were like, 'No. No, no!' They didn't want to spend six hours doing a load of laundry. They wanted to have a machine, push a button, and have their clean laundry."

Doug maintained a gas generator that provided limited electricity at night, and powered communication by CB and VHF radio. If the rain was too heavy, even radio traffic was unintelligible. In this hermit existence, Doug was overjoyed. He sent hundreds of gushing letters to colleagues and friends back in San Francisco. He sounded evangelical as he sang the praises of his analog existence. As they attempted to decipher his radical new lifestyle, his friends asked one another the same question: "Has Doug gone mad?"

Chapter 9

Stalking Tigers in Siberia

The world is only as big as we allow it to be. Wild places and animals pass along their secrets only if we listen. A touch of danger would help. You need to know you can die: a surprise rapids the size of Lava Falls, a bad stretch of black ice across an ice chute, a whiteout on a glacier, or maybe a bear or, especially, a tiger.

—DOUG PEACOCK

In the early 1990s, as Russia opened for capitalist development, a unique economic boom occurred—a tumultuous period that many observers compared to the lawless Wild West. As the legions of staid, gray bureaucrats were overrun by a crop of freshly minted oligarchs, Yvon Chouinard snagged an invitation to visit a little-known Russian wilderness area. It was one of the last remaining habitats to observe wild Amur tigers, and Yvon was eager to explore. There were rumors the forest would soon be clear-cut. Tompkins jumped aboard the trip, also excited to explore the wild coastlines and thick forests. Hyundai and International Paper were plotting to buy the old-growth forest and cut it down, a source in Russia told them. "Maybe, if we like it, we'll buy it," Tompkins declared.

Jib Ellison had experience running rafting trips behind the Iron Curtain and was among the few adventurers with the street smarts to snag highly coveted travel permission. Ellison applied for hunting visas; this would raise little suspicion, and though he knew his group would travel far outside the limited permission given to hunters, it would allow them to legally enter Russia with expedition gear.

Ellison, Tompkins, and Chouinard were eyeing an area in the Russian Far East, near the Sea of Japan, which their contacts in Russia described as a wilderness cornucopia flush with roaming pods of wild boar, spotted leopards in the trees, Asian brown bears scraping off the bark below, and prowling the forest floor the world's largest and most powerful wild cat—the Amur tiger.

Ellison was a river rat. Years earlier he had founded RAFT—"Russians and Americans For Teamwork"—and taken travelers into the then–Soviet Union. He'd met Tompkins at a slide show about RAFT at San Francisco's famous Zuni Café, and Tompkins later accompanied Jib exploring the USSR in kayaks and rafts.

On that previous trip to the USSR, Jib had saved Doug's life. On the verge of descending an unknown river, Doug insisted that they raft down a boulder-clogged river with Class V whitewater rapids. Jib resisted. Although he was among the youngest member of that expedition, he was insistent that the descent was too dangerous. After a heated discussion, Doug's idea was rejected. Several days later, on a helicopter overflight, they passed above the rapids and it was clear that Jib had saved them all from certain death.

Doug called his fellow adventurers "The Do Boys." It was a phrase he filched from a Japanese comic book with terrible translations, and the moniker fit. At the heart of these Do Boys trips was a deep faith in individual self-reliance. There was no expedition leader. "It's like, saddle your own bronc," explained Ellison. "It's your job to look out for yourself. And if you abide by that ethos, and people take it seriously, it's ultra-responsibility, right? Nobody is in charge. It's much more old-school. Essentially, it's 'don't fuck it up.'"

The Russian coastline along the Sea of Japan held dozens of protected inlets that served for Russian Navy bases. Vladivostok held a strategic submarine base, while the Sakhalin Peninsula was home to ICBM missile launch sites. From the beginning of the Cold War in the late 1940s the area had been largely off limits to US citizens for the ensuing sixty years. Soviet strategic military zones didn't interest Tompkins; he wanted to explore the larger neighborhood—the lower reaches of the Bikin forest stretched across 1,500 square miles, making it the world's largest untouched forest of broad-leaved cedar and the only natural corridor through which the Russian and Chinese populations of wild tiger bred.

Ellison arranged logistics for the trip as the Do Boys needed flights, estimated transit times, and travel dates. Doug was free for six weeks, but others in the group had only three weeks. The Do Boys could arrive in June, Jib explained to their contact in Russia. "That makes no sense," insisted the Russian. Tiger studies were always conducted in winter, he pointed out. The snow made it possible to track and study the giant cats. Jib said he understood, but the Do Boys were busy and finding a common date was difficult. June was the only three-week window when everybody could go.

"That's fine," the tiger expert chuckled. "We will make sure there is somebody in camp to meet you. But if you *do* see a tiger? It will be the last thing you ever see. You don't go looking for tigers in the summer, because the grass is five feet high—everywhere." Jib went back to the Do Boys and repeated the message: "The tiger biologist says that if we see a tiger it's going to be the last thing we see, because it's going to be eating us." Jib never forgot the response. "The Do Boys were all like, 'Well . . . that's when we can go. Let's just go anyway.'"

In June 1992, Doug, Jib, and Doug Peacock boarded an Aeroflot flight from Seattle. Landing in Khabarovsk, Tompkins, Ellison, and Peacock were met by a representative of Greenpeace who smuggled them via train to Sovietskaya Gavan, a highly restricted city where nuclear subs were based. They were almost certainly the first Americans to visit in years.

They arrived late and went immediately to meet with three mafiosos in a dark basement with a single hanging bulb for light. The Russians smoked and tried to shake down the Americans who needed a boat. The conversation went explosive several times until a vastly reduced price was agreed upon. They would use a boat from the Red Cross.

Tompkins, Ellison, and Peacock had ten days to explore before their other friends arrived, so they rode the boat down the coastline to explore a nature reserve. When the men lowered their kayaks into the Russian river it felt like they were paddling into a new world. "The reserve here is unlike American parks with our administrators and wilderness visitors; nobody's out here," Peacock wrote in his journal. "More mammal species live in these woods than any place I've ever traveled. The Russians leave these lands to fend for themselves, protected by remoteness. The wild corner of my soul is envious."

Peacock and Tompkins had radically divergent lifestyles, yet were aligned in their love of the wild. Peacock had camped with grizzly bears in the wild national parks of the western United States. Tompkins had just spent twenty years running a multinational corporation, yet in wilderness, they both found an antidote to urban ills. Modern society was a sickness, they agreed, and wilderness experience the remedy. They both worshipped beauty, and Peacock knew firsthand the face of evil.

As a special forces medic for the US Army, Lt. Peacock had been awarded two Bronze Stars for his courage in combat. He returned to the United States in 1968 amid the grisly revelations of the My Lai massacre. Peacock's psyche was marked by events in the Vietnam highlands. He had left a slice of his soul in Vietnam. Flashbacks to the deaths of Vietnamese civilians, friends killed in combat, being shot at too many times to count—it all haunted him. Back in the US, Peacock found human civilization unbearable. To tame his demons, he migrated into the wild. He camped in Utah, then in Montana, where he lived alongside wild grizzly bears. A PBS documentary entitled "Peacock's War" showcased his bravery and singular love of the wild.

Peacock maintained a deep friendship with the naturalist author Edward Abbey, who took Peacock's campfire tales, changed the name to protect the guilty, and converted him into George Hayduke, a fictional ex–Green Beret "wilderness avenger" and hero of his novel *The Monkey Wrench Gang*. In Abbey's eco-warrior fantasy, Peacock is a saboteur, wreaking vengeance on bulldozers and dams alike. Peacock was so close that after Abbey died, it was Peacock who helped leave his body in the desert to be recycled by nature, his corpse eaten by wild birds, and then his bones chewed up by animals.

After chugging down the coast, the men camped near a remote village. In the evening as they ate by a campfire near the riverbank, they met a local hunter named Gorbachev. They hired him on the spot. Who better to track wildlife? Gorbachev took them to his homestead and offered a local delicacy: moose heart, fresh baked. The scene was out of a movie. His clothesline sagged with hundreds of dried fish. Children smoked the fish over a campfire wearing boots and shoes made from animal skins sewn so that the fur was on the inside. After sunset, the men gathered. Drinking was the common language.

Peacock wrote up the scene when the hangover finally wore off. "Tompkins proposes yet another toast from the end of the table: 'Screw Hyundai!' I watch him pour the contents of his glass into the planter he has chosen to conveniently sit next to, while I gag down my poison," wrote Peacock. "Doug does these toasts five more times, five more big shots of vodka, which he dumps out when no one is looking. I pray he doesn't kill the plant and hope I don't offend my new friends by puking on the table. Tompkins has a big, shit-eating grin on his face."

Entering the forest, Tompkins scouted the terrain. But within forty-eight hours the Americans were forced to abandon camp. They were being threatened with expulsion for not having proper visas. Local gangsters extorted them, demanding huge sums to keep their visit secret. The three explorers abandoned the forest in a rush. Tompkins was crestfallen. In his trip diary Peacock jotted, "This man loves forests."

Returning to Khabarovsk, they picked up their three fellow expedition mates: Yvon Chouinard, Rick Ridgeway, and the relative newbie Tom Brokaw. They were now joined by Dimitri, a biologist they had met at the nature preserve and whom they had convinced to join them. The Do Boys worked via an amorphous sense of leadership. Their credo was "learn on the run." They didn't slow down or teach rookies. "Doug taught me to kayak. He basically put me in a boat and said, 'Follow us,'" recounted Chouinard. "Same thing with Brokaw. We took him on some climbs without any experience—he couldn't even tie a knot. But he was tough."

A lifelong outdoorsman, Brokaw was known to millions of Americans as the host of *NBC Evening News.* As chief political correspondent for America's leading TV network, Brokaw broadcast live as the Berlin Wall fell, and held the first interview by a US network with Mikhail Gorbachev. On their first day in Khabarovsk, Brokaw was also the first to shout "foul" when KGB officers tried to shake down the Do Boys with a $2,000 demand for a "tourism permit." Brokaw ignored all diplomatic protocol and bellowed "This is banditry!"

Facing a shakedown from the secret police force, the Do Boys huddled. Was it time to ditch the handlers? The decision was unanimous: they would sneak off on a rogue helicopter flight. "We gave a helicopter pilot a bunch of Marlboro cigarettes. The pilot was stoked. We threw in some *Playboy* magazines and off we went," Peacock recalled. "The KGB wouldn't know where we were going."

Inside the helicopter, the Do Boys celebrated as they sat alongside Russian military men. The rusted helicopter felt overloaded and decrepit. Peacock, who carried years of experience patching together

Battered by a storm, clinging to an icy wall with no ropes, Doug Tompkins makes a daring climb in Scotland on a trip with best friend Yvon Chouinard.
Credit: Yvon Chouinard

Doug Tompkins (*left*) with Olympic ski medalist Billy Kidd (*right*) on a break from training at Portillo Ski area in Chile, 1961. Tompkins and Kidd "borrowed" the motorcycles from a BMW dealership for a test ride that lasted until a police alert was issued to apprehend the two *gringos locos*. *Credit: Tompkins Family Archives*

Doug Tompkins founded The North Face in San Francisco, 1964. His first mail order catalog (shown here) promoted the idea that climbing and outdoor life were best enjoyed by packing minimal equipment, traveling light, and living as close to the wild as possible. *Credit: Tompkins Family Archives*

The North Face store in San Francisco attracted a crowd, including singers Janis Joplin and Joan Baez. The opening of its 1966 winter season included an in-store concert by the Grateful Dead. As security, Tompkins hired an infamous motorcycle gang, the Hell's Angels. *Credit: Suki Hill*

Four friends calling themselves "The Fun Hogs" preparing to leave Los Angeles in 1968 and drive 16,000 miles to climb Mount Fitz Roy in Patagonia. Doug Tompkins (*right*) prepares to leave with (*right to left*) Dick Dorworth, Yvon Chouinard, and Lito Tejada-Flores. *Credit: Patagonia Archives*

Best friends Doug Tompkins (*left*) and Yvon Chouinard (*right*) on the road during their 1968 road trip to climb Mount Fitz Roy, an icy peak only two teams had summited. *Credit: Lito Tejada Flores*

Mount Fitz Roy became visible to the climbers after months of exploring and driving. Chouinard would lead the challenging climb up the dominant middle spire. Snow and blasting wind made camping impossible. On the mountain, they planned to live in ice caves. *Credit: Doug Tompkins*

Trapped for weeks in an ice cave meant to be a temporary refuge, Tompkins and his fellow expeditioners nearly ran out of food as eighty-mile-an-hour winds threatened to blow their shelter apart. In the photo Dick Dorworth (*right*) wrote in his journal alongside Lito Tejada Flores (*left*). *Credit: Chris Jones*

Doug Tompkins became a trailblazing business leader in thc early 1970s. From a San Francisco apartment, he helped build what would eventually become a billion-dollar empire. *Credit: Suki Hill*

Susie and Doug Tompkins (with their first daughter) worked tirelessly to build Plain Jane, their small clothing startup, into a worldwide brand sold in hundreds of retail outlets. *Credit: Suki Hill*

Doug Tompkins with Susie and their daughters, Quincey and Summer, as they prepared to fly in a small plane from California to South America. The parents removed the backseat and put in foam cushions for the children to sleep and play on. *Credit: Paul Ryan*

"Image Director" Doug Tompkins supervising a photoshoot for the Esprit catalog. Using employees and "real people" as models, he rejected the idea of paying professionals. "Anyone can hire Brooke Shields," he quipped. *Credit: Helie Robertson*

Design and architecture magazines praised the Esprit style as innovative, daring, and exciting. Much credit was given to "Chairman Doug," featured here in one of the many cover stories dedicated to explaining the Esprit phenomenon.
Credit: Sharon Risedorph / Blueprint

Doug Tompkins (in kayak) being lowered over a waterfall by friends. A lifelong adventurer, he spent three or four months each year in the wild. He joked that it was his MBA—Management By Absence. *Credit: Rob Lesser*

Kayaking replaced climbing as Doug Tompkins aged. A first descent of a river in Chile nearly got him shot as he inadvertently trespassed the private grounds of dictator Augusto Pinochet. *Credit: Rob Lesser*

Flying above Patagonia, Doug Tompkins felt free. During his more than twenty years in the region, he flew some seven thousand hours above the mountains and rivers, allowing him to memorize one of the most spectacular wild areas still remaining on Earth. *Credit: Barbara Cushman Rowell*

Clear-cut forest along the border of a US national forest in Washington State. Flying above these brutal clear-cuts motivated Tompkins to abandon his life as a fashion tycoon and focus on halting the destruction.
Credit: Daniel Dancer

Doug Tompkins at home in rural Patagonia, dressed impeccably. Living off the grid in the wild, he spent months brainstorming ways to slow logging, sabotage hydroelectric dams, and halt the construction of environmentally destructive megaprojects. *Credit: Jo Schwartz*

Rural Patagonia is so windy that a gust can easily flip over a small airplane. Doug Tompkins knew to tie down his plane after landing in a pasture. As a pilot, he was widely praised. One friend opined that he had "quick, fighter-pilot stuff." *Credit: Galen Rowell*

The wildlands of Patagonia held irresistible attraction to Doug Tompkins. Many areas were accessible only by small boats or aircraft and, pelted by some nine feet of rain a year, held few human settlements. *Credit: Antonio Vizcaino*

Doug Tompkins (*left*) at the campfire with their guide, NBC anchor Tom Brokaw (*back right*), and Jib Ellison (*right*) as they explored the Russian Far East on an expedition to track tigers. *Credit: Rick Ridgeway*

Yvon Chouinard (*left*) hiking with Doug Tompkins. Best friends for nearly a half century, they often hiked, climbed, and surfed together. Back in civilization, they worked tirelessly to promote the protection of wild and scenic areas around the world. *Credit: Rick Ridgeway*

In his dream to save vast swaths of South America, Doug Tompkins began mapping areas that he could turn into nature sanctuaries. Everyone laughed and told him it was "Mission Impossible." He took that as a challenge. *Credit: Gary Braasch*

Tompkins viewed the power of multinationals as so dangerous that he spent millions on newspaper advertisements to warn the populace. He called corporations "The Invisible Government." Such advertising campaigns spurred on the antiglobalization movement. *Credit: Doug Tompkins*

Wild puma returned to their native haunts and thrived as Doug and Kris Tompkins implemented a conservation strategy known as rewilding. *Credit: Chantal Henderson*

Volunteers worked for years to remove some five hundred miles of fencing inside Chacabuco Valley in preparation for enticing native animals to return as part of the creation of Patagonia National Park. *Credit: Patagonia Company Volunteers*

The hyperactive and fabulously coordinated Doug Tompkins kayaked and hiked through remote areas of Patagonia as he designed strategies to create a string of national parks. It would be his final brand and his grand legacy: the Route of Parks. When it was finished, it would even be visible from space. *Credit: Beth Wald*

In an effort to stop plans for a hydroelectric dam project on a dozen wild rivers, Tompkins helped organize and fund a pro-Patagonia campaign to save the rivers. He paid for billboards promoting the slogan "Patagonia Without Dams." *Credit: Doug Tompkins*

The endangered macaw parrot barely existed in northwest Argentina when Tompkins arrived in 1998. Using animal trainers and veterinarians to teach injured and domesticated parrots how to fly required years of study. To bring back the wild, Doug and Kris Tompkins had to learn how to "rewild" the land. *Credit: Beth Wald*

Doug Tompkins (*center*) working alongside Argentine cowboys in traditional garb. *Credit: Tompkins Conservation*

Doug Tompkins (*right*) in Antarctica, aboard the ship *Steve Irwin* as it chased a Japanese whaling fleet in an effort to stop the slaughter of some six hundred whales. Tompkins bought the fuel for the mission (costing a quarter million dollars), which was run by the Sea Shepherd Conservation Society. *Credit: Eric Cheng*

Reliving his days as a baseball player, Doug Tompkins practices throwing stink bombs onto the decks of Japanese whaling ships during environmental protests off Antarctica's coast. *Credit: Eric Cheng*

Chilean cowboys protest a $3 billion dam project in Patagonia. By highlighting the beauty of the region and its local culture, Tompkins and a broad coalition of activists helped provoke a surge in regional identity that drowned the would-be dam in bad publicity. *Credit: Linde Waidhoffer*

Doug and Kris Tompkins often worked twelve-hour days, stuck behind their computers, but whenever possible they found relief by hiking in the hills. Wild puma that ranged nearby made the walks especially exciting. "I began to rewild my mind!" said Kris.
Credit: Rick Ridgeway

battlefield casualties in the Vietnam War, wrote that it felt like being stuck inside "a flying boxcar."

Free of handlers, the Do Boys looked nervously at the rows of metal jugs sloshing with milk, which only added to their worries that the chopper was dangerously overweight. There were no seats, and the helicopter regularly stopped in villages to drop off milk. It was hot and the windows were open, but their helicopter escape had worked. "I thought we'd never be seen again," said Brokaw. "We got out of there in a manner that was totally illegal."

The men now celebrated the first goal of every Do Boys journey—to abandon civilization with glee. Flying low over the forests, Tompkins looked out the window and studied the woodlands below with the eyes of a pilot. He had seen this wilderness on maps. This was a vast, uncharted, green territory, sliced through by birch forests, wild rivers, and small clans of pioneers living in frontier villages scattered over an area the size of Portugal.

Dimitri, the biologist who they had invited along, studied their only topographic map as the helicopter rattled above a sea of trees. They were searching for a place to enter the Bikin River. What was the most challenging entry point? Could they begin on the uppermost section of the Bikin? Peacock asked for the map, moved to the window and instantly the map whooshed out the porthole. Peacock stared in disbelief. In each hand, he gripped a third of the map. The portion in the middle showing the upper Bikin was missing. "Everybody started high-fiving and saying, *This is a much better trip*," recounted Brokaw. "The Russian guy was laughing. He said, 'I know where the headwaters are. We can find that. Rivers always flow downstream.' Doug could not have been happier. We were out there on the edge, not knowing where we were or where we were going."

Landing in a field near the river, the Do Boys set up a campsite along the riverbank and explored the forest. The men saw wild boar. Scratches on trees were signs that bears lived nearby, and they heard what sounded like wolves baying at night. Human settlements were few and far between. "There's a jungle tom-tom working," noted Brokaw, "because when we pulled into these tiny, tiny temporary camps for hunters and gatherers, they knew we were coming."

Camping in the front yard of a rustic family, the Do Boys shared a bathroom with their hosts, bathed in the freezing river, smothered local bread with fresh honey, and wolfed down a smorgasbord of smoked fish as trappers paddled past with furs on their way to market.

Camping by the water's edge, the Do Boys found a pool of pungent tiger urine. Paw prints a half foot across suggested this was a young male tiger, a dominant predator that left scratches and marked its territory every few hundred yards. With few settlements for hundreds of miles, the Amur tiger ruled this forest. The Do Boys knew that walking through the tall grass where tigers roamed was dangerous, so they strapped on masks. One Do Boy wore a Ronald Reagan mask, while another donned a mockup of George Bush. They spun the masks to the back of their heads, hoping to ward off marauding tigers. "We'd heard that farmers in India put a mask on the back of their heads," said Chouinard. "Tigers will only attack from the back, not from the front. You can stare them down. So, we bought these masks."

Walking in the forest, Brokaw was convinced the tiger was straight ahead of them. "It wasn't very far away, it was clear. The odor from the urine and the whole sense that the tiger was here was very evident to us. We were out of our minds," said the newscaster, "because in the next year or two years after that, there were these accounts of this marauding tiger wiping out hunters, and settlers, and everybody else."

Suddenly, Dimitri freezes and motions me forward: a tiger track glistens in the mud. The track in the wallow appears to be only about a day old, around five inches across. The print, Dimitri says, of a young but dominant (about five years old) male cat that has replaced the previous dominant male cat, who was killed by a poacher. The young tiger leaves scrape marks every few hundred meters and spray-scents on territorial tree markers. We stop at such a tree. The bark has been rubbed off by Asian black bears that also are attracted to the strong scent. I get down on my knees and press my nose against the bare trunk. The pungent fetor of tiger fills my nostrils and—for just a second—I travel with the big cat, orange and black stripes flashing barely perceptibly through the sea of green, undulating cone fern, into the wild and predatory world that not so very long ago was my own.

—DOUG PEACOCK

Kayaking in the remote woods, sleeping in tents, sharing stories around the campfire, and enjoying a can of beans heated in the fire was paradise for the Do Boys. Beyond the edge of civilization, they found a connection to a lifestyle they each needed. The tiger trip was a celebration of the wild. Out in the woods, life slowed to a natural rhythm. Running water was their soundtrack.

The friends floated down the river, quiet as they listened for bears and fished for a dinner and then cooked together. Tompkins often broke the silence, annoying the others. "Doug would paddle behind me," said Brokaw. "He would be beating my eardrums about *You're not doing enough on radical environmentalism*, and I finally turned around and said, *Doug, goddammit, I'm out here to get away from that, not have more of it in my ear.* He said, *Okay, okay. I get it.* Then ten minutes later he started up again."

As they wrapped the kayak expedition, the men needed a ride back to civilization. Their only option was a daily train that passed near the forest, late at night. The 11:00 p.m. train had sold out nine months earlier, but Russian fixers had arrangements with the train crew that meant, ticket or no ticket, the highest-paying customers rode the train.

The Do Boys paid up, then disguised themselves with handsewn clothes and thick jackets, and followed precise instructions: when the train stops, throw the gear aboard. Under no circumstances ask for permission. Don't speak English. Don't worry about the feisty babushka with gold teeth, she was part of the team. The Do Boys boarded the train without tickets, swapping furtive glances and communicating in rudimentary sign language. The ruse worked and they arrived by morning in the city, where their guide Dimitri had insisted that the Americans come to meet his family, living nearby in Vladivostok.

Foreigners were not allowed into Vladivostok. Under Russian law it was a "restricted military city" and off limits to foreigners, and especially journalists. "Brokaw went crazy. He thought he was going to be arrested in a highly publicized incident," laughed Jib Ellison. Dimitri knew the local greasers and fixers who maintained the rusty state apparatus, so he took his guests to a flea market and bought each man a new outfit. Every Do Boy was given a Russian winter fur hat so large it could be pulled down, past his eyes. "Just pretend you are kind of drunk," Dimitri insisted.

Sneaking into Vladivostok without visas, Dimitri took his clandestine visitors to his tiny little apartment where they feasted with the family, sang folk songs, and collapsed asleep on the floor. When they recovered from their vodka-forged friendship, they visited the offices of the Pacific Institute, and went on a sightseeing jaunt through an area off limits to foreigners and even to many Russians.

"There's some old Russian submarine pulled into the port," recalled Ridgeway. "It's a weekend, pedestrians everywhere. And the submarine is open. Dimitri said, 'Perfect, we're going to tour the submarine!' Brokaw is really shitting at this point, because we're now touring a submarine in a city we're not supposed to be in, dressed like Russians. Of course, Tompkins thinks this is perfect. He is in heaven, because we are breaking every rule."

Tompkins was exhilarated from his time in the wild, but he was certain that Russia was *not* a place to invest. Though enchanted by the thriving forests, he wouldn't be buying one. The corrupt practices of post-Soviet capitalism involved bribes, mafia, and too many dubious characters. Tompkins knew that forest land in South America was just as cheap, and in Chile, he believed, the locals played by the rules.

Chapter 10

Two Exotic Birds in a Strange Land

A traveler to Patagonia, a region of legendary wildness and beauty, will find the experience transformative. Here is a landscape of unparalleled natural treasures, one of the last great strongholds of wild nature.

—DOUG TOMPKINS

Three weeks camping in the Siberian wilderness reignited the passion inside Tompkins. He returned to Chile more determined to build his own shrine to nature—a 500,000-acre park in the wilds of Patagonia. "This will be the largest private nature reserve in the world," he told his young assistant Daniel Gonzalez. The opportunity was fleeting, he stressed, as the forces of development were flattening forests in British Columbia, Madagascar, Indonesia, and Brazil. Many of Chile's native forests were already gone, he warned.

Surveying the maps of southern Chile, Tompkins imagined ways to "monkey wrench" the lumber and chip industries who planned to buy the forest. He was certain he could never stop the forces of industrial development, but he was equally convinced he could buy time.

Tompkins helped Gonzalez build a command center. Like pioneering explorers, they gathered every map they could find as they sketched their plans for a massive park across these unknown lands. They were given free use of a small office above the Esprit store in Santiago. Juan Enrique Abadie, the local Esprit concessionaire, adored Tompkins and was delighted to share resources with the brand's legendary founder. In their small office Gonzalez and Tompkins imagined Eden. The apex predator in these lands were muscular mountain lions known as pumas. They weighed up to 120 pounds and feasted on native deer and imported sheep. Tompkins sketched out a park half the size of Yosemite. He understood the beauty of asking for the impossible. He named his plan Pumalín Park and said it would be the largest private conservation initiative in the world.

Rick Klein, who had been the spark for Doug's initial interest in Chilean forestry conservation, watched as Tompkins appropriated it. Klein's wild idea of protecting ancient forests in Patagonia was blooming, and, even though Doug left him completely on the sidelines, he was ecstatic. Klein jokingly called Pumalín Park "Doug's Ego System."

Tompkins thrived on the logistical challenges. Why not put together a park, he strategized, by buying land from absentee landowners holding title to untouched forests? Like the first descent of an unknown whitewater rapid or kayaking over a waterfall without knowing what lay below, the idea thrilled him. As his vision expanded, so too the wall of maps in his makeshift office, where Tompkins hung a sign that echoed his edicts from the Esprit HQ—*No Detail Is Small.*

When Gonzalez found a property for sale, he gave Tompkins the coordinates, and the two climbed into a small plane, flew down, and explored. Tompkins was now an experienced pilot with thousands of flight hours. He felt comfortable as he landed on country roads and bounced across remote cow pastures. Once on the ground, Tompkins rustled up a local cowboy, loaded packhorses, stuffed food and sleeping bags in a burlap sack, and rode into the backcountry on a horseback trip that might last two days, or maybe six.

"He could go so extreme, so comfortably. He didn't care that life had changed radically. I think he had had enough of the other life," said Marci Rudolph, who had also made a radical switch from working at Esprit in San Francisco to living in Patagonia. "I couldn't believe that Doug, someone who loved eating out at nice restaurants and who thought that San Francisco was the greatest city in the world, and who traveled all over the world for business, now is hunkering down in the south," she recalled. "He didn't care about nice champagne; he didn't care about food. Everything switched. We did salad plus fish, if one of the guys caught one, or if not, it's *let's open a can of tuna.* . . . I don't think Doug changed. I don't think Doug became a different person. I think he took all that energy he had and put it into Chile."

Tompkins imagined Pumalín Park having concentric rings of protection. The terrain was so rough and the rainfall so persistent that few humans even tried to live in the zone. Keeping humans out was not difficult; in fact, it might be more of a challenge to entice like-minded settlers to live in the rainy Reñihue watershed. Doug described to Marci his plans to design a pioneer village, a low-tech (and at times no-tech) manual-labor-based economy where friendly neighbors collectively built fences, cleared away stumps, and devoted their lives to sustainable agriculture practices that cultivated, rather than mutilated, the land near the proposed park.

Tompkins prioritized wild animals and native flora but was practical enough to realize that small, human settlements were his best chance to create a long-term buffer zone. He imagined settling a family in each major watershed—they could live off the land and monitor the environmental health of the ecosystem. "We'll teach bee-keeping and how to market the honey," Tompkins proclaimed in a letter to colleagues back in the states. "The idea is to encourage things that break them out of the cycle of poverty, get them over that desperate line of survival."

As his land holdings grew across Chile, stretching from the Pacific Ocean all the way up the Andes to the border with Argentina, Tompkins owned land that chopped Chile in two halves. Tompkins viewed this as a random by-product of his larger conservation dreams. But Chilean authorities felt the nation's sovereignty was threatened. In a costly tactical error, Tompkins blithely assured admirals and generals not to worry, he was going to gift all the land back to the Chilean government for free. He just needed more time to put together his massive plans for a national park, Tompkins assured the befuddled military leaders.

In a country emerging from the dark lies and manipulations of a secretive military government, few believed that an altruistic gringo had arrived to invest $150 million and then deliberately gift it all away, no strings attached. Chile had been deprived of nearly every Earth Day celebration from 1973 to 1990 as military curfew and repression kept a lid on most forms of social organizing. Activists had been specifically targeted. By the early 1990s environmental awareness was just emerging, and repeated promises by Tompkins to save the forests and protect the deer felt like a weak cover story for something far more nefarious.

"I don't think anybody believed him," remarked General Juan Emilio Cheyre, commander in chief of the Chilean Army at the time. "No one believed him because it was so strange. What do you mean, he's going to donate the lands? Wasn't it the other way around? Wasn't he just buying more land? It was a verbal promise, outside of protocol, and it wasn't something with a firm date, with details and a handover plan. There was no real notion that he would really donate the lands."

Tompkins was ignorant of the decades-long history of property battles in the region where he'd set up. Villagers were long accustomed to being swindled, abused, and run off the land by Chile's ruling aristocracy—often working in cahoots with powerful politicians. "When we came into that region, there was already a deep frustration from *colonos* [early settlers] toward the government and toward what they perceived as outsiders," said Gonzalez, who traveled often with Doug into the wild to meet with locals. "For thirty years they had been promised title to their lands, ownership to their lands, and nothing had happened. We were just one more actor in what was already a frustrating situation for locals. In many ways, we were perceived as a threat."

"No one was against his conservation plans. Who could be against a person who wants to preserve nature, protect forests and waterways?" suggested Hector Muñoz, then a high-level official in the Ministry of the Interior. "The problem was the magnitude of his philosophy. We found out he was a follower of Deep Ecology, a believer in biocentrism, which means that nature comes first, and that humans are just another element of nature and that all beings have the same rights. He even argued with me that rocks have the same rights as human beings! That's when we began to understand his ideological character; this man was absolutely convinced of his beliefs. He had an incredible fanaticism, and deeply held convictions."

Tompkins was oblivious to the enmity his conservation plans generated. Fresh from upending the world of teen fashion and flush with cash, he spent little time pondering what might go wrong. Looking forward, he was the first to admit there was no master plan. He also knew that to pull off a dream like creating the world's largest private park, he needed a team. At Esprit his successes were facilitated by a small group of invisible allies—a loyal cadre including his exceptional business manager, Debbie Ryker, and Dolly Ma, his longtime logistics manager who, like an air traffic controller, sorted a steady stream of incoming ideas. Ryker and Ma managed Tompkins's post-Esprit financial affairs from a small office in San Francisco that functioned as a watchdog operation for his wild ideas. But what he most lacked was a dedicated co-conspirator. And it was his best friend, Yvon Chouinard, who introduced him to the woman who became the love of his life.

In early 1990, Yvon Chouinard needed a break. For two decades he had guided the Patagonia clothing company with a passion similar to the enthusiasm he brought to ice-climbing ascents and kayaking adventures. Patagonia's guarantee of durable, well-designed clothing had earned the trust of a vast marketplace of loyal customers. Annual sales soared from $20 million in 1985 to $100 million in 1990. But Patagonia was no cash cow. The company reinvested profits in building a stronger company and also donated millions to grassroots environmental groups fighting to protect their community.

Chouinard needed to transfer day-to-day operations to a new management team, so he flew them to the tiny village of El Calafate, not far from where he had been trapped in a snow cave with Tompkins thirty-three years earlier on the expedition that had inspired him to name his company Patagonia. In El Calafate, he knew he could evoke a love of the wild that so inspired not just the company's name but its commitment to environmental activism.

Kris McDivitt, Patagonia's CEO, was leading the transition effort for the company. The thirty-seven-year-old was a former ski racer. Lithe, athletic, and determined, Kris carried the physique and energy of a long-distance runner. She was one of the original seven Patagonia employees, and as longtime CEO had built the Patagonia brand while inspiring a deep loyalty from customers and employees that was the envy of the industry.

On the last day of the Patagonia management retreat, Doug flew over the Andes, landed in El Calafate, and settled in for lunch with his best friend. When Doug arrived, Yvon happened to be sitting next to Kris. The Patagonia Inc. team had finished their informal agenda and were preparing for a flight back to California the next day. Doug sat down next to Kris in a rustic backwater restaurant, slapped her on the back, and proposed, "Hey, kiddo. How are you doing?" He quipped, "Why don't you fly back to the United States with me?"

Kris refused the offer. Doug insisted. "You're just going to fly back commercially," he teased.

When she continued to reject his advances, he asked if she could bring a bag of books back to California for him. She agreed, thinking it was a small kayak bag. Doug came back with a sixty-pound duffel bag of books. As she later recalled, "We both knew that the bag of books was really the motivation to be in touch again."

Kris viewed Doug as a maverick, a brilliant visionary, and the best friend of her lifelong boss Yvon. Kris had bumped into Doug over the years. Her former husband had climbed with him in the Trango Tower area of northern Pakistan, and she was well aware of his reputation both as a playboy and as a charismatic business leader.

Kris was one of the few people in the world who knew Yvon as well as Doug. A native of Southern California, she began packing orders and handling paperwork with Yvon when she was just a seventeen-year-old. From her beginnings working for Yvon in the mailroom, back at the climbing equipment company that was known as Chouinard Equipment and when she was known as a "surfer girl," Kris showed a talent for building teams. She was brash, self-confident, and loyal to Yvon. She steadily rose higher in the company as the company expanded. Her brother, Roger McDivitt, became CEO.

When Roger left the company, Kris was told to take over. Suddenly and by accident, Kris was in charge of Patagonia. Given her absolute lack of management experience, she cold-called bankers in the Ventura area, explaining, "I've been given this company to run, I don't want to destroy it, can you help me figure out what to do?" Her unorthodox manner and charm worked. Quickly she took the reins of Patagonia. Her uncanny marketing and leadership within the growing Patagonia team quickly led to her becoming a trusted executive.

Her friendship with Yvon helped filter the ever-changing ideas he'd bring back to Ventura. Kris knew how to implement the good ideas and ignore the wacky ones, which (over time) led Yvon to refine his more outlandish ideas. They were a dynamic team, and among the employees of the privately held corporation Kris was the only one ever given stock and partial ownership of Patagonia.

Kris earned a reputation as an adroit businesswoman with a knack for handling the details of Yvon's sprawling empire and covering for his extended absences. She laughed at herself and cultivated a sense of humor. When Chouinard insisted that she buy for the company a bizarre shoe style he'd found in Japan, known as Reef Walkers, McDivitt refused. When he insisted, she wrote in dark ink on the beams above her desk: "My boss made me buy 20,000 pairs of Reef Walkers." And she had Chouinard sign the ceiling. The shoes never sold, but the incident, and the graffiti, were etched into company lore.

For weeks after their chance meeting in Argentina, Doug pursued Kris. Repeatedly he invited her to dinner. She diplomatically swatted away all his entreaties. Doug tried repeatedly to convince Kris to come up from Ventura but she was reluctant. And engaged. Their schedules overlapped, however, long enough for dinner at Doug's San Francisco house. Entering his home, she was impressed by the art, the style, and the food. "He cooked pasta for me with this amazing red sauce—probably the same sauce he served to all the women in his life!—but then we stayed up all night talking, and I left early the next morning."

Despite numerous invitations to mandolin concerts and environmental lectures, Kris held back. But when Doug heard that she was headed to Paris, he flew there. Inviting Kris to a night on the town, he took her to an antiglobalization teach-in. Kris noted that Doug was in the inner circle and held longstanding friendships with several key participants. After the workshop they went to dinner, then strolled Paris by night. At 3:00 a.m., Kris announced she had to go. The next day she was headed to climb Mont Blanc. Sitting down in the taxi, Doug stalled. He asked her to go to Chile with him. "He said, 'I will never let anything happen to you,'" Kris recounted. "The way he said it was something else again."

When Kris told their mutual friend Jerry Mander she was planning to take up Doug's offer and visit him for twelve days in his remote cabin in Chilean Patagonia, Mander laughed and told her, "No one can last twelve days in a cabin with Doug Tompkins."

Kris ignored the advice. "I decided, what the hell? Go. Just go. No expectations."

Arriving in the cabin in rain-swept Patagonia, Kris wondered what she'd done to her life. When Doug asked how she was doing, Kris spilled her guts. "How do you *think* I am doing! I've just dropped an atomic bomb into my life."

Doug took it all in and then asked, "Can I tell people you are my girlfriend?"

Neither of them ever looked back. With Doug, Kris felt like she'd found her authentic self. It was if her other life was a facsimile. Fighting for nature with this intense and brilliant partner felt right. Doug's friends were stunned. "She opened him up a lot. And yeah, he was completely different after he met Kris, that's for sure," said Chouinard. "He was more bombastic before. He was, *It is my way or no way.* He would argue everything, didn't see anyone else's viewpoints. A lot of people did not get along with him because he was too self-centered and too sure his opinion was the only one. She was able to control him and really change him around."

Kristine McDivitt and Douglas Tompkins were married in San Francisco at a simple ceremony in the San Francisco City Hall in 1994. As witnesses, Doug brought Jerry Mander, the progressive PR pro he'd worked with since the early Grateful Dead concert at his North Face store, along with his wife, Elizabeth. After the wedding champagne, they went home to finish designing a book, then held a meeting with Earth First! founder Dave Foreman, whom they'd left alone for a couple of hours while they slipped out for the low-key wedding ceremony.

Later, in the remote cabin that was now their base in the wilds of Patagonia, Kris found herself scrubbing the mold off apple trees, reading by candlelight, and deciphering Spanish language maps of the region. They only had electricity for a few hours at night when the generators were fired up, and the rains pounded day after day on the roof of their home. Despite all predictions that she would feel claustrophobic and trapped, Kris felt protected. The home felt like a nest. "The transformation of Doug when they fell in love and Kris joined him down there, he was really happy; he felt strengthened by Kris," said friend Peter Buckley. "I think he had just that much more capacity to do what he wanted to be doing. Kris was the perfect complement to his skills. Doug was just missing that gene for hospitality and taking care of people. They pushed each other, and Kris was fiercely supportive of and loyal to Doug. And she would stand up to him. He could be kind of rough or a curmudgeon. And Kris would call him on his bullshit."

Together in their remote farm, so far off the grid that they used CB radios to talk and rainwater to shower, Doug and Kris launched a relationship overflowing with enthusiasm and a sense that they had stumbled on a once-in-a-lifetime opportunity. They had a mangled farm surrounded by hundreds of thousands of acres of wildlands, and a desire to change the world. What would they paint on this half-million-acre canvas?

Chapter 11

Salmon Wars

National parks have a lot of benefits. One is, they get people out into nature. They are a form of social equity, as they belong to everyone. Parks provide to all citizens a chance to renew the spirit in a stressful world. Parklands provide a place for reflection and contemplation. Each time a new park is created anywhere in the world, it will help society understand the deep necessity to share the planet with other creatures, and preserve biodiversity.

—DOUG TOMPKINS

Living on their rainy farm, Kris and Doug often invited Yvon and Malinda Chouinard down to explore their new homestead. One morning, Doug and Yvon went kayaking in the bay near a sea lion rookery on a rocky outcropping an hour's paddle away. Doug rowed up to what looked like a floating log five feet long and semisubmerged in the coastal estuary. When he got closer, Tompkins was struck by a foul odor, then a bitter understanding. Bloated and filled with air, the "log" was a dead sea lion. Inspecting the carcass, Doug probed his fingers into the fur until he located a puncture wound—a smooth entry hole. He suspected it was from a high-caliber bullet, since the salmon farmers were known to hire sharpshooters to drill a bullet into the animals when they came up for air. It was part of their brutal strategy to keep the sea lions from eating their prized salmon.

Tompkins hauled the unwieldy carcass over the bow of his kayak and, balancing his rancid cargo, rowed toward Fiordo Blanco, a salmon farm granted a concession on land, together with the rights to install an industrial salmon farming operation in the fjord, in the middle of what Doug assumed was his front yard. "He throws that dead seal on one of their boats and he says, *I got a reward for anybody with any information on who killed this seal*," said Yvon.

Using his cash to pry out the truth, Tompkins offered one million Chilean pesos for information about the sea lion killings. "It's two times anybody's annual salary. So, it was war," said Imhoff, the environmental desk manager who worked with Tompkins at Esprit. "He was putting his money out there and he was fighting for the critters, for the creatures of the world. And for that, I give him all the credit in the world. But he paid for it." Without realizing, Tompkins had just fired the first salvo in what was soon to be an undeclared war with some of Chile's most reactionary forces. Some of whom worked for the salmon industry.

Under Chilean law, the navy controlled certain rights to the shoreline. Owners of the Fiordo Blanco salmon hatchery were able to get government permits to build pens offshore as well as a beachhead operations center on land. Lit up at night with klieg lights like a high school football stadium, the salmon farm was noisy, smelly, and dirty. Dump trucks began depositing salmon guts, salmon heads, and miscellaneous offal on Doug and Kris's property.

The odor, the mess, and the grotesque nature of 60,000 pounds of salmon remains dumped at sea and washing ashore sent Tompkins on the legal warpath. Tompkins called Pedro Pablo Gutierrez. Not only was Gutierrez his attorney who had helped on the original paperwork to purchase Reñihue but he also functioned as the Spanish equivalent of consigliere. As a young partner at Carey & Carey, the law firm negotiating the most important and prestigious foreign investment projects, Gutierrez worked in the same offices as those who managed the family fortunes of the dozen or so families that dominated the narrow Chilean economic landscape. Under insistence from Tompkins, he filed a lawsuit against the salmon farm, citing the Chilean constitutional guarantee to live in an environment free of pollution.

Back at Reñihue farm, Tompkins ordered his team to padlock the fences along the roads, thus marooning the salmon farmers on the tip of the peninsula. Despite the run-ins with Tompkins and local fishermen and a long list of labor violations at their processing plants, the farmed-salmon business brought a welcome dose of steady, year-round work to the region. Jobs in the area were generally low wage and often seasonal.

When Tompkins declared that he opposed all the salmon operations, he was anything but subtle. He rattled off a dozen reasons to implement a moratorium on salmon farming concessions, especially as practiced at Fiordo Blanco. These rustic operations left his pristine bays bobbing with plastics and the coastline littered with tarps, Styrofoam, and abandoned buoys. And then there was his critique of the fish—engorged creatures manufactured down to the specific pink hues of their inner tissue, which came from dyes added to the salmon food pellets. Like a homeowner choosing the color for a bathroom wall, salmon producers in Chile selected their favorite shade of pink from a half-dozen options, ranging from the fiery glow of a sunset to a more subtle grapefruit tone. Ocean activists were appalled that the mighty salmon had migrated so far down the food chain, and they derided farmed salmon as "chicken of the sea" and a mash of mushy protein no tastier than "oatmeal with gills."

Marine biologists warned that overcrowded salmon pens were prime breeding grounds for epidemics, including the ISA virus and an invasion of parasites that munched away the outer layer of the fish in what looked like an attack by piranha but was actually a locust-like plague known as "sea lice." The industry responded to the outbreak by effectively marinating the live fish in antibiotics. Tests showed that

Chilean salmon held thousands of times more antibiotics per pound than salmon farmed in Scotland or Norway.

As annual revenue from the salmon industry jumped into the hundreds of millions of dollars, pristine lakes in Patagonia were smothered by salmon feces, and oxygen levels in the lakes plummeted. The ensuing public outrage kicked the salmon operations out of the lakes and back into sheltered ocean bays. Soon these bays began to die off. The industry denied the damage. Tompkins bought a remote control underwater robot and hired a pilot to conduct an undersea survey. The submarine-like craft documented the damage. Foot after linear foot of uneaten salmon food, salmon feces, and garbage completely altered the ocean floor. "They did an environmental assessment," said Chouinard. "And their conclusion was that it was a dead zone—there was nothing alive down there. And that was because of the salmon farm."

I heard once from a Canadian fish farmer in our fjord in Chile that he could give a shit (his words!) about the outbreak of viruses that would leave some survivors who had a genetic disposition to resist the disease and that they would use those fish's genes to produce the GMO fish that his company was developing. I knew I was in front of evil, cold evil at that. It was a critical moment for me from an ethical point of view, that cemented a resolve in my own mind about the whole human project, and I never forgot that. The guy was cold-blooded serious and never had a trace of regret that wild salmon would be gone, just as long as it suited his business objectives. He did not need to tell me that without wild salmon, fish farming markets would be the only ones left.

—DOUG TOMPKINS

The escaped salmon were also an environmental hazard as they went on a conquest of southern Chile—invading local rivers and streams,

Reñihué Homestead

1995

Pumalín Park
Feral Cows
Hobbit House
Rio Negro
Hangar
Airstrip
Guest House
Garden
Reñihué Schoolhouse
Tompkins House
Gardens
Gardener House
Milking Shed
Caretaker House
Family House
Family House
Reñihué Fjord
Palena House
Carpentry Workshop
Mechanical Workshop
E
S
N
W

devouring the larvae and eggs of native species, and rapidly establishing themselves as apex exterminators. So many salmon were escaping from pens that traditional fishermen in the region netted more and more of this surprisingly valuable invader with flesh the color of a tangerine. Salmon even flung themselves across roadways near Puerto Montt. Drivers watched agog as ten-pound salmon swam, leapt, and flopped across a flooded street in a misguided migration. But these were not wild salmon, argued the Chilean salmon association, they were fugitives and runaways.

Using a law designed to allow ranchers to reclaim lost cattle, the association drafted laws to prosecute and punish local fishermen who attempted to sell *any* salmon, whether caught in a river or in the open ocean. In a controversial effort to strangle competition against their monopoly, salmon producers fought to make the artisanal sale of noncompany fillets illegal. The lobby group argued that as all "wild" salmon came from company stock, the producers therefore maintained and owned the rights to all their offspring. Thus, a black market for salmon emerged. Drably dressed, middle-aged men started to troll Puerto Montt, their Styrofoam coolers of "illegal fish" atop weak-wheeled hand trollies. Zigzagging through the downtown, they half-whispered their street-dealer pitch—"salmon, salmon, fresh salmon."

Despite pressure from Tompkins, the security force from Fiordo Blanco continued sniping at sea lions in a futile campaign to dissuade the hyper-persistent animals from breaking and entering. Tompkins did not realize that his submarine patrols and "reward poster" had ignited a firestorm of opposition, or that Rene Patricio Quilhot, an employee of the salmon farm, was a retired army colonel and ranking member of the DINA, General Pinochet's secret police. Accused of killing a Spanish diplomat and torturing others during the dictatorship, Quilhot maintained connections to the military and intelligence service in Santiago. Like Tompkins, he'd settled at the ends of the Earth in rural Patagonia. Unlike Tompkins, Quilhot owned a phone.

He made some calls. Quilhot and Fiordo Blanco wanted *el gringo* off their back.

It wasn't long before the powers from the central government in Santiago set up a special committee. Although it was never so stated, their mission was clear: convert the life of Doug Tompkins into a bureaucratic hell. They began to torture him with the small levers of power operated from inside key government ministries, what the French disparage as the *petit bureaucrate*. "After that, we started having constant friction with the most powerful people in Chile, the owners of these industries," said Carolina Morgado, a whitewater rafting operator Tompkins had hired as his personal assistant to implement his conservation strategies. "In the offices we had in Puerto Montt," she recollected, "we got graffiti that said *'Muerte a Tompkins'*—Death to Tompkins."

Fueled by a pliant right-wing press in Santiago, a cauldron of rumors was unleashed against the Tompkinses. Miguel Serrano, a Chilean novelist and a loud Nazi sympathizer, organized leafletting campaigns in southern Chile with flyers announcing that Kris and Doug Tompkins were secretly planning a Jewish homeland, even though they were both raised as Anglicans. Then a rumor spread that Doug was bringing in bison. American buffalo would soon be marauding in herds and replacing the cows. So many calls came into their office denigrating Doug Tompkins as a "dirty Jew" that the receptionist took to screaming, "He's not a Jew, he's a shrew!" alluding to his stingy nature and using the Chilean phrase *"No es judio! Es jodido!"*

Others suggested Señor Tompkins was looking for land to bury nuclear waste for the US. Or was he breeding a super mountain lion that would wipe out cattle, sheep, and even a way of life? "I wanted to be out of the limelight; instead I got sucked right into it," said Tompkins. "If your project is big, it's automatically controversial. I was advised to have public relations people running things, but I don't believe that's a good solution. That puts a layer between you and reality. It doesn't give the public a genuine impression of who you are. I should have taken two months of total-immersion Spanish. I was not as articulate as I could have been in the Chilean media."

The smears against Tompkins followed the ideological vein pushed by Chile's neoliberal president, Eduardo Frei. Son of a former president, young Frei didn't follow in his father's footsteps right away. First, he became a hydraulic engineer. Critics joked that when Frei looked at a pristine river he saw flowing electricity. As president, Eduardo Frei was determined to open up southern Chile's vast resources to billion-dollar schemes. He supported destroying native forest lands in order to replant with more commercial eucalyptus and pine. He supported damming pristine rivers to power a huge aluminum smelter. Projects such as these were not seen as billion-dollar boondoggles but rather as signs of progress. The Frei government's prodevelopment agenda courted international investment and sought to distance the country from the shadow of the Pinochet human rights abuses while maintaining key tenets of the dictator's free-market orthodoxy.

With spray-painted death threats coming from Nazi kooks and government officials eager to take down his conservation plans, the harassment campaign against Tompkins picked up. Doug and Kris consulted with Gutierrez, their lawyer, who had a clear understanding of power politics in the government. They decided to hire a retired detective from the Chilean PDI, the Chilean equivalent of the FBI. Gutierrez reported that the detective had found tidbits of a plot. "There was a plan to plant drugs in Doug's apartment in Puerto Montt and then to have a police raid on the apartment. And to say, *Oh, we have discovered drugs and this explains everything. Mr. Tompkins is a drug dealer*."

The next step of the plan entailed expelling Tompkins from Chile. Problem solved. But Gabriel Guerra Mondragon, the ambassador of the United States to Chile, raised hell. He emphasized to the Chileans that Tompkins had invested in Chile and followed the rules of the game. He had brought the money through the approved capital routes and invested accordingly. Any insinuations that Mr. Tompkins was a drug dealer were ridiculous.

Suspicious breaches of private information about Tompkins then leaked to the press, suggesting that they were being spied on. A team of electronics specialists was even flown to Doug and Kris's office in Puerto Montt. Yes, they told Tompkins, your phones are tapped. Determined to react, Doug asked what they could do. "We sent a charge back the other way, and blew out the tapping system," said Kris. "And it blew out our entire phone system."

"We call it 'the weight of the night,'" said Enrique Correa, Chile's most influential lobbyist, as he explained the hidden hand of fascism used to threaten and muzzle Chile's emerging democracy in the mid-1990s. Correa described the repressive power as concentrated both in the Chilean military (where, even after the return to democracy, Augusto Pinochet was named commander in chief, then designated "Senator for Life") and in the conservative wing of the Catholic Church. The conservative Catholic movement known as Opus Dei was so strong in Santiago that church leaders were able to label the heavy metal band Iron Maiden as "Satan worshippers" and even shut down the band's planned 1992 concert in Santiago.

Sifting through one of Foundation for Deep Ecology's statements, the Opus Dei congressional delegation unearthed his defense of a woman's right to abortion. They went on the warpath, attacking him as a promoter of abortion. For the extreme Catholic faith, Tompkins was also guilty of a second sin—a failure to put homo sapiens atop all of God's creations. With his Deep Ecology philosophy and millions in cash, Tompkins was, in the eyes of powerful Chilean officials like Frei, a dangerous renegade.

Gutierrez, the lawyer working with Doug, recognized that Chile was controlled by an economic mafia. He also knew that Chileans jousted with staplers and legal briefs more than stilettos or bullets. From his high-level perch, Gutierrez had the altitude and eyesight to narrow the list of suspects. Who was coordinating this ugly campaign? He suspected that Belisario Velasco, Chile's vice minister of the interior and a heralded democracy activist who'd waged a clandestine campaign against the Pinochet dictatorship, was the architect of the anti-Tompkins sentiment. Velasco, said Gutierrez, "was going after Doug on a daily basis in the press."

Determined to calm the waters, Gutierrez phoned Velasco to suggest that rather than battle via headlines and press campaigns, they simply meet and hash out whatever had the minister in a dither. "I went there and we had a meeting. . . . One of the things he mentioned in the meeting was, 'Look, you know that the DEA is worried about Mr. Tompkins, and you know if this leaks to the press, this is the end of the man in Chile.'"

As he listened to the veiled threat to frame Tompkins that the powerful minister so calmly described, Gutierrez prepared his counterattack. "I said, 'Look, Mr. Minister, Mr. Tompkins does not even drink mineral water. He does not drink coffee. Indeed, he does not consume any kind of drugs and he has nothing to do with drugs whatsoever.'" said Gutierrez. When Gutierrez related the latest threat against Tompkins to Ambassador Guerra Mondragon, who as the highest ranking US official in country was legally the head of the DEA in Chile, "The ambassador started calling Velasco every week, just sort of checking in, letting him know we were watching."

Velasco then planned an ambush to smear Tompkins. First, he invited the American to his office in the presidential palace to discuss the American's extensive landholdings. During the brief encounter, Velasco asked about the properties Tompkins held in Argentina. But Tompkins dismissed any notion that he owned land contiguous along the borders of both countries. He felt confident that he'd put the matter to rest, noting with ease that his Argentine lands were hundreds of miles away. But Velasco pounced. The following day he held a press conference and denounced Tompkins. "He has admitted that he has lands on both sides of the border," said Velasco in effect, bringing to the forefront the idea that the conservation lands threatened Chile's sovereignty.

Tompkins was naive about the approaching maelstrom. Only after months of ambushes and political sabotage to his conservation plans did he truly calculate the dimensions of the smear campaign. "We had no idea of the dark manipulations of Velasco, which became more evident over time," Tompkins said. "So, we began to deal with these situations, seriously distracting us from the daily work of clarifying land titles, doing other negotiations, placing infrastructure, personnel, and everything else. Many things were left aside or simply thrown away."

"I was operating with a presidential mandate," said Velasco when asked about his role in Le Affaire Tompkins. "One time when I was meeting with him in person he said, 'Chileans should be thankful for the forest I'm preserving.' And I replied, 'That forest has been there for a thousand years, and no one is going to exploit it, because it's too steep, and the machines can't get in there to work.'"

Tompkins's lawyers worked overtime to squelch the dirty tricks, wasting countless hours as tax authorities demanded reams of documents with so little logic that it appeared they were on a fishing expedition. Gutierrez dug into the root of the problem, asking one mid-level tax supervisor why Doug was being given just forty-eight hours to respond with tax papers. "Whose instructions are you following?" he asked. "The woman in charge, who was very nice, said to me, 'Look, I will look aside, and you just read the memo I have in front of me.' And it was an instruction from the undersecretary [Velasco] to go after Mr. Tompkins and catch him on anything."

"Velasco had a personal dislike for Doug as a person, and he was the vindictive type. If Doug was a superhero, this was the supervillain," said Gonzalez, the young aide. "I had no experience with diplomacy or any of that. And I would just say in the meetings, 'Well, this is what's going on. And here are the maps.' And he hated that. At the time we thought the best course of action was just to be honest and open. Over the years, you learn that it doesn't work that way."

Kris was buckling under the stress. "I thought the world was crashing down," she said. In long discussions Malinda Chouinard encouraged Kris and explained that conservation battles were often fraught with controversy. Malinda cited a book that detailed the creation of Grand Teton National Park in Wyoming. "That was the first time I realized we were just part of an ongoing story," said Kris. "Wherever conservation took place, this kind of animosity, this conflict, was inherent to the process."

Then an infiltrator stole internal documents from the local Tompkins office. "This guy Carlos Martinez came to us because he wanted to interview Doug for a big Chilean magazine, as a journalist," said Morgado, the assistant working with Kris and Doug. "One night I was coming back to my office to check something, and I found him in my office photocopying papers. 'What are you doing in my office?' I asked. He tried to give a reply but couldn't."

Soon the stolen papers and documents from the various Tompkins-led foundations appeared in press accounts lambasting environmental activists, and also surfaced later in the hands of a Chilean congressional committee investigating Tompkins. "Based on all this knowledge that we managed to get from Carlos and other sources, we established the threat that Tompkins represented," said Hector Muñoz, chief of staff to Velasco, in a tacit admission that the Ministry of the Interior had received the stolen information.

"There was something messianic about him," said Carlos Martinez, the alleged document thief. "He felt he was saving a piece of the world. 'It's me against the world. Development is bad, progress is bad, there shouldn't be anything here.'" Martinez, the infiltrator, had spent time preparing his role by flying with Tompkins and posing as a journalist as he picked Doug's brain while they soared for hours over the land, Doug all the while chronicling the destruction of Patagonia. "What seemed negative to me about Tompkins," said Martinez, "is that he would be controlling an important part of the country, and what he planned for it was a type of dam against progress and development."

"Political opponents have all kinds of dirty tricks and we see that every day," Tompkins wrote to his colleagues. "We don't know whether we can weave our way through all those things or not; we'll just have to see. We take it day by day and try to use the best strategies and tactics possible to keep our project alive and going through. They face a terrible dilemma in trying to do something really drastic against us."

The harassment continued with low-flying airplanes buzzing the Reñihue homestead, at times just above the rooftops, and with cameras hung out the window to snap surveillance photos. Sergio Cardenas, an agent with the Chilean Agencia Nacional de Intelligencia described the harassment mission as nearly a full-time job. Cardenas flew in the planes to surveil Tompkins and described the ease with which the state harassed the renegade foreigner. "It was free to punch him," said Cardenas in an interview. "So the government punched him, some members of Congress punched him. Those on the right punched him imagining that he was coming with Jews, and for appropriating territory, and those on the left punched him for being a gringo," explained Cardenas. "Both Minister Belisario Velasco and President Frei understood that punching Tompkins was free; it had no associated cost."

A reporter asked Tompkins how he was handling the attacks. He explained that after realizing he was being ambushed, he joined the fray and he deliberately provoked fiery public debates and used controversy to grab the attention of government officials and the general public. Said Tompkins, "From my point of view, what is happening in Chile, as in many other places, is due to groups like ours that create a debate that attracts the attention of the government and the public in general. We should be proud of taking action and being at the vanguard of this process. . . . These things are like every environmental battle everywhere: the forces of conservation vs. the forces of development. We're kind of like lone soldiers fighting it out."

Yet the harassment went beyond a spirited public debate. When trees fell after big storms and washed down nearby rivers, Tompkins had his carpenters set up a temporary sawmill and use the fallen wood for building material. But when a journalist photographed the operation, two days later the local newspaper accused Doug of cutting down native trees. The headline was: "Tompkins is cutting down native trees to build." "That was our daily bread," said one collaborator. "People were always looking for a way to do a smear campaign."

"It was curious because you had a lot of business guys and entrepreneurs who were coming to open forestry companies and cut the trees. And there was no objection to that," said Hernan Mladinic, a sociologist who had worked at a Chilean NGO fighting the export of native trees as wood chips. Tompkins hired Mladinic to build relationships into the highest echelons of La Moneda, the Chilean presidential palace. The reception Mladinic received was cold: "This guy who was saying he was going to promote conservation, why should Chile tolerate this guy?" Another problem, Mladinic recognized, was that Doug Tompkins was deliberately outspoken. He was not a quiet conservationist. As Mladinic recalled, "Doug always told me that making a park is the easy part. Although there is controversy, usually people looked at it in a positive way. But when you start criticizing the development model? The logic of industrialization, of industrial agriculture, and technology? These things are dear for certain groups of people."

Despite the opposition, the distraction, and the threats, Tompkins continued his land purchases. He bought entire valleys, purchased volcanoes, acquired stands of old-growth trees. With substantial financial contributions from key allies in the US and Europe, Kris and Doug built a broad base of donors. Doug alone had spent $15 million buying land. But still a single key piece was missing: a 75,000-acre parcel smack in the middle of the proposed Pumalín Park. The missing piece was owned by a Catholic university in Chile's main port, Valparaiso, and the university leadership wanted to sell the lands known as Huinay. The university was in dire financial straits and eager to unload the remote, unwieldy tract of forest. Tompkins met with the leaders of the university and explained his plans for conservation and environmental protection. They listened carefully and then asked about his stand on abortion. Was it true, the bishop queried, that Doug was forcing his workers to have abortions? Was his Deep Ecology belief a cover for antihuman beliefs? The arch conservative Catholic monsignor Jorge Medina initially agreed in private to sell Huinay to Tompkins, then abruptly made a public declaration that he was undecided. A week later, the archbishop of Puerto Montt, Bernardo Cazzaro, attacked Tompkins's beliefs because they "put nature's interests above those of humans."

"The Huinay property was owned by the Catholic University of Valparaiso," recalled Velasco, the powerful minister. "And the director was a friend of mine for many years, Bernardo Donoso. I told him, 'You're not selling that property to Tompkins. Either the Chilean government is going to buy it, or we will find a company that will.'" The university ended up selling it to ENDESA, a politically connected energy conglomerate with close ties to President Frei. Velasco crowed over his victory, commenting, "Tompkins nearly had a heart attack, because it divided his lands in the middle, and he wanted an uninterrupted tract."

In ENDESA, Velasco found a powerful utility company willing to finance the government's dirty work. With no viable business motive for purchasing forest lands, ENDESA, in essence, bankrolled the harassment plans developed by President Frei and his advisers in the Chilean presidential palace. ENDESA spent millions to buy a remote forest for the sole purpose of hindering Tompkins's park plans. When his master plan succeeded, Velasco mocked Tompkins in the press. "Tompkins has no reason to feel threatened just because they decided to buy a piece of land in the middle of what he was planning as a park," said Velasco with irony and glee. "He'll just have to scale down his ambition; his park can be 270,000 hectares instead of 300,000."

Tompkins was irate. He lambasted President Frei "as just an engineer" and categorized the government operations against him as "diabolic." He declared, "I don't have time for this kind of show. I don't deserve this kind of treatment by the authorities." Huddling with his lawyers in Santiago, Tompkins considered the unthinkable: Had he lost the battle?

Part III

Chapter 12

The Land of Shining Waters

In the long term, the economy and the environment are the same. If it is unenvironmental, it is uneconomical. That is the rule of nature.

—MOLLIE H. BEATTIE

As Chilean authorities surveilled, harassed, and threatened Tompkins, on the other side of the Andes Mountains a group of Argentine conservationists looked on with curiosity at this gringo with cash and a love for nature. Opinion polls regularly found that Argentina's populace held some of the most anti-US sentiments of any nation on Earth, yet the country's conservationists admired Tompkins keenly. If his conservation plans weren't all going to work in Chile, they wondered, would he give Argentina a shot? The invitation arrived from Argentine National Parks director Francisco Erize. Would Señor Tompkins be interested in touring potential conservation sites in Argentina? A list of biodiversity hotspots was delivered in a tone of collaboration distinct from the confrontations in Chile.

Doug and Kris flew to Buenos Aires and then embarked on a whirlwind tour of Argentina. They visited a range of damaged but biologically valuable ecosystems. They viewed rain forests and glaciers and were offered the opportunity to save either whales or condors. Penguins or parrots. Argentina—the sixth-largest nation on Earth when measured by landmass—held a vast breadth of flora and fauna that was under attack by modern agriculture, forestry, and fisheries practices.

As they toured Argentina's ravaged ecosystems, Doug and Kris were aghast at the damage done by cows. Grilled steak was the national dish, beef exports were a source of billions in foreign revenue, and Sunday BBQs were a national pastime rivaling football. The native ecosystems were being decimated to clear space for cattle, and the government was lobbying the US to increase beef imports fivefold, from 20 to 100 tons a year. The Argentine government even appointed the tennis star Gabriela Sabatini as "beef ambassador" as part of a lobbying campaign. In Argentina, if it wasn't cow, it wasn't "meat." When foreign travelers asked waiters for the vegetarian option, they were served chicken.

As they flew back from the Yungas region in northwest Argentina, Kris and Doug stopped in the Corrientes province in northeast Argentina, where the Paraná River fed a vast tributary system near the borders with Brazil and Paraguay. The province of Corrientes was home to wetlands the size of the Florida Everglades, with floating islands populated by capybara, caimans, eagles, and egrets. A century-long trade in feathers and fur, as well as more recent massive monoculture pine tree farms, had driven many species to near-extinction.

The floating islands were thick with poisonous snakes, dangerous sinkholes, and water pools teeming with piranhas. The tangled geography included dry grasslands, savannah, forests, and marsh. The unique mix of land and water sheltered pockets of biodiversity, including bands of orange howler monkeys and the nearly extinct pampas deer. Jaguar had been hunted to extinction in the zone and the maned wolf whittled down to a tiny population. "We landed there, and I really thought we'd landed in hell," said Kris, describing her initial impressions of the wetlands known as Iberá. "It was flat and hot and buggy, and I didn't see anything I liked or recognized and wanted out of there. But Doug did."

In Argentina the national parks were a source of national pride and, unlike in Chile, the Argentine parks were a significant source of revenue for the government. Doug realized that the legal framework for national parks was exceptionally strong in Argentina. Park fees might be siphoned along a dozen different illegal paths, but the parks themselves, he estimated, would be neither stolen nor sold.

Four months after first landing on the remote airstrip in Iberá, Doug bought San Alonso, a 26,000-acre island in the middle of the wetlands. The island provided a strategic vantage point. Like a lighthouse watchman, Tompkins now had an observation post. His property had enough pasture for an airstrip, so by land and air, Tompkins photographed and documented illegal intrusions into the swamplands. He envisioned turning the entire ecosystem into a single park, and he figured the fight could never be rougher than dealing with the Chileans.

I think that more and more leading businessmen are deciding these conservationists don't know what they're doing. They don't know how to manage, because running a park is basically like running a business, and they can do it better. That's why I think Doug and Kris have been so successful—they are businesspeople. They're not tree huggers. That makes a big difference. We need more people like that, no doubt. . . . People say I'm hypercritical of conservationists around the world and in particular conservation organizations or governments and I always say—and I truly believe—that Kris and Doug are some of the greatest conservationists that have ever existed. They are in a class all by themselves. What they've accomplished in their lifetimes just from a conservation perspective, which started relatively late. Just think about if there were 500 individuals like them or 1,000 individuals like them on earth! The planet would be a completely different place. And there are not, there just aren't. There are a couple of handfuls of people like them on earth, which is sad. But it's spectacular what people can do if they set their mind to it.

—J. MICHAEL FAY

Iberá was rugged, at times lawless, and not a place where an arrogant *yanqui* could invest without first understanding the local landscape. Seven years of battling Chileans on the other side of the Andes had taught Doug to study the local culture, then act. He immediately went flying with a local pilot named Carlos, a bad boy and daredevil who could guide him around. Carlos was notorious for flying upside down like a stuntman. Doug and Carlos fed off each other. Doug could fly as wild as the local hotshot, if not better, and Carlos pulled out his revolver and blasted away at the sky as they careened down the runaway en route to yet another flyover of the wetlands. The association with Carlos was both a natural inclination and a shrewd move by Tompkins. Regardless of how locals interpreted his presence, there was no chance anyone would characterize the new *yanqui* as a "normal guy."

The gregarious and feisty Correntinos were the opposite of Chileans who tended to be introverted, suspicious, and decidedly nonviolent. "Corrientes felt like a warrior province. That's how it imagines itself, and it's not a joke," said Kris. "They're tough, very independent, and many did not like the national government, and they don't take to foreigners."

Iberá was also home to a mythical figure—the *gaucho*. But this was not the noble gaucho figure of southern Argentina. In these northern wetlands, the word "gaucho" also meant a fugitive, a solitary man who at times abandoned society to live in the bush and perhaps survives as a hunter-gather hermit.

The marshland's soundtrack was a raucous squawk of birds. Yellow cardinals and white woodpeckers flew through the reeds and pockets of forest that provided a home to more than 300 species of birds—nearly as many species as found in all of Europe. Flamingos and geese migrated from Patagonia every year to pass the winter. "Iberá gets migratory birds from the north during one-half of the year and from the south in the other half of the year," explained Juan Ramon Diaz, a wildlife photographer working in Iberá.

As part of the larger Paraná watershed, the Iberá marshlands were once riverbeds. When the great Paraná meandered north, the wetlands filled with rainwater year-round and created an entrance to the Guarani Aquifer, the world's second-largest aquifer. The difficulty in traversing the area had converted the wetlands into a cauldron of myths, rumors, and conspiracy theories. As much as they distrusted *yanquis* from the United States, however, there was one enemy even more repugnant to locals—the political bosses back in Buenos Aires. Correntinos felt abandoned by the politicians in the capital and the rest of the federal government. Their flag was Argentine but their loyalties provincial. "They say Corrientes is another country," said one local. "And that they give you a passport when you come here, and if Argentina gets into a war, Corrientes will support her." Tompkins found Iberá's rebel attitude comforting. In a land without rules, regulations, or routines, anything was possible.

Doug realized that in Iberá wetlands he had found a diamond in the rough. "Like a painting," he told friends as he described the view from flying above the wetlands that native Guarani culture named Iberá, which when translated means "The Land of Shining Waters."

In 2000, Kris sold her Patagonia company stock and founded Conservacíon Patagónica, a California nonprofit designed to facilitate the creation of national parks. She investigated areas rich in biodiversity and in Argentina explored the country from its northern jungles, where jaguars still prowled along the border with Brazil, to the southern extremes near the Argentine lakes region. Together with Doug she mapped out the fragments of once-thriving biological corridors. Were there remnants enough left to stitch together a critical mass? If not, were individual islands of biodiversity worth saving?

Kris and Doug debated which ecosystems to prioritize. Local environmental activists had compiled a depressingly rich list of finalists.

Did they want to first protect native forests from soybean plantations? Or was it more important to go offshore and develop Marine Protected Areas (known as MPAs) to allow the ocean's remaining fish and marine mammals a bit of protection after three centuries of plundering?

"They were a team," said the environmental historian Harold Glasser. "Doug was the sort of stalwart bulldog, but she was the one who knew how people work and smoothed things over and knew how to bring out the best in people. Kris is the one who always seemed quite rational, really open-minded, and a very good critical thinker."

Finally, Kris put her effort (and stock proceeds) into Monte Léon, an overgrazed, eroded, coastal ecosystem used and abused as a sheep ranch along the Atlantic Coast in southern Argentina. Monte Léon stretched for 165,000 land acres including twenty-five miles of wild coastline, penguin colonies, and a fragile, desertlike coastal vegetation nibbled to nothing by thousands of sheep.

Working with Erize, the esteemed Argentine National Parks executive director, they laid the groundwork for a three-way transfer. Conservación Patagónica would fund the purchase of the ragged ranchlands through a nonprofit called Fundación Vida Silvestres Argentina, which in turn would transfer the lands to the National Parks Administration. In thirty-six months the project was complete, and Monte Léon was added to the Argentina national park system. It was the country's first coastal national park and it opened the path for Marine Protected Areas to create ocean-based conservation initiatives in the future. "When you go to these anti-American Argentines and explain that you have all this money and that you want to buy private lands, fix them up, and then donate them back to the state, they don't have a hard time finding that to be a great idea," said Tompkins. "You are paying the costs, then allowing them to renationalize the land."

Kris and Doug Tompkins were now establishing a yearly routine, a seasonal migration, from Pumalín to the Iberá, where they were becoming ever more involved in a campaign to protect the wetlands. Winter in Patagonia was damp and dark at just the time the Iberá wetlands felt spring-like. Journeying from their rainy Reñihue ranch to Argentina was a pilgrimage that Kris likened to "a honeymoon over and over." During the days long drive, they took turns reading aloud to one another. "We would just have a great time laughing and debating things in the car," said Kris. "I would get tired of sitting, so he would stop and read, and I would run out in front of the car and then signal to him when I'd gone far enough. He'd put whatever he was reading aside and come get me."

Expanding their footprint in Iberá, Doug and Kris hired Sofia Heinonen, a pioneering Argentine biologist with seventeen years' experience in the Argentine National Park Administration. Heinonen, a confident and passionate advocate for nature, was fearless, yet even she was stunned by Doug's audacity. His introductory tour and chat with locals (which Sofia had suggested) became a revolutionary declaration. Tompkins explained matter-of-factly that he had come to build a national park, reintroduce the jaguar, and reinvent the local economy, with conservation as the backbone.

Heinonen was flabbergasted. She watched in bewilderment as Tompkins publicly announced his revolution to one rural community after another as they traveled around the perimeter of the vast wetlands. She had been working for Tompkins just four weeks, and already his meetings with the various conservative agricultural and farming associations had sparked an uproar. Tompkins presented his vision with such confidence and certainty that the locals took a deep swallow, unsure how to react. "It was like he had said, *I'm going to the moon, after that to Mars, and then I'm going to build my house on Jupiter*," laughed Heinonen. "They thought he was crazy, that it was an impossible dream, like he was an extraterrestrial that didn't know where he had landed."

To better help him understand the unique biology of the region, Tompkins also hired Ignacio Jimenez, a Spanish wildlife biologist. But before he could delve into the intricacies of rewilding the wetlands with giant anteaters, pampas deer, and jaguar, Jimenez had to deal with a rebellion on his front step as he attempted to quell fears by locals that Tompkins was a fraud. "For the Argentines it was like, *No way; tell me another story because I don't believe it,*" said Jimenez. "First, every American wants to screw us; second, every millionaire is an asshole; so there must be a hidden agenda here. It was shocking for me. There was so much aggression, so much hatred. Any private company owned by a foreigner that bought land to do business as usual, let's say mining or forestry or rice—it wouldn't create any controversy. But it was the fact that a foreigner was buying lots of land for *conservation*. That created the controversy. He was a UFO. Since nobody had seen this before, they had to create alternative versions of the story because they would never believe the real story."

Among the critics was Argentine senator Sergio Flinta, who represented the Iberá region in the provincial legislature. "The first objection I had was that he is an American coming to buy land. We've always had a strong *yanqui*-phobia here," he said. "The second was that he was coming for the water in Iberá, and that he was coming to create a barrier so that Iberá could not be utilized by the province. He was interested in conservation over production, and there were things being said about him—that in Chile he had cut the country in two. This generated a strong opposition in the local farmers federations."

Tompkins recognized that the limited access to the wetlands played in his favor. Few ranchers or even their farmhands felt comfortable trekking into the mysterious swamp that was known as the exclusive territory of the *mariscadores*—nomadic hunters who hunted and captured anything they could sell. *Mariscadores* trapped otter, used salt licks to lure marsh deer, and sent up smoke signals from island to island as they moved cattle herds through the wetlands. They captured so many *caimans* that the population plunged from hundreds of thousands to mere hundreds.

"Iberá was an empty space. It had no fauna when Doug bought it. But he saw the chance to fill it with critters, because the habitat was there, and was not being used for something else," said Heinonen. "When Doug flew over Iberá he saw the opportunity to revive it with wildlife, and to try rewilding as a tool on an empty stage."

The yellow anaconda, which reached lengths of eighteen feet, had been poached into near-extinction as the *mariscadores* earned a few dollars selling skins. Egret feathers kept entire families employed as they used sticky traps to glue the feet of the birds to branches, then harvested them for a few ounces of hat feathers. The crash of such garish fashion led to a steep decline in the *mariscador* population. From a total of several thousand, their numbers fell to perhaps only 300—which included a roaming population of outlaws.

"They're not necessarily dangerous bandits. It's more like they don't understand what law is; for them it doesn't exist," explained Heinonen. "For example, a *mariscador* goes to town for a party, meets an underage girl, and brings her back with him to the marshlands. He didn't know how old the girl was or why, afterwards, there was a legal complaint lodged against him in the town—which he now can't go back to anymore. He doesn't understand why they would be looking for him."

To better understand who and what was living on the floating islands, Heinonen offered to investigate. Saddling up, sometimes with her horse dragging a canoe, she ventured far into the swamplands. When the water got too deep, she tied up the horse and used a pole to push her wooden canoe through the reeds. For ten months she explored the hinterlands, spotting endangered marsh deer and inspecting scat to understand what species were still living in the wilds. Her mission was to survey the hermitlike encampments.

Making contact with these remote homesteads, Heinonen came across a wide zoology of humankind. Fugitive murderers. Kidnapped women. Illicit families with children born and raised in the swamp. Sometimes she would find a solitary man whose only companions were dogs and horses. Sitting by campfires she tried to decipher the mystery of the wetlands by listening to stories of the olden days when pelts of jaguar could be sold at the local market and guns trumped government. She gained their confidence by drinking *mate* tea with them.

Mate drinking is a ritual all across Argentina, Uruguay, Patagonia, and parts of Paraguay and Brazil. Dried stems and leaves of the yerba mate plant are stuffed into a cup made out of a squash gourd, then soaked in water. After adding sugar (or not) the tea-like beverage is sucked up through a thick, stainless-steel straw. The yerba mate plant is native to Corrientes, and few rituals more clearly distinguished city dweller from country folk. Moving the straw causes debris to clog it. Stirring with the straw was akin to insulting local culture and considered a colossal blunder.

Heinonen evoked surprise and respect as she ventured into the backwaters. Unaccompanied women rarely explored these quarters. "It was a life without resources—only really old guys, or completely undocumented guys outside of the system, who didn't speak Spanish, only Guarani," said Heinonen. "We hired many of those guys to be park rangers and thus no more hunting. I didn't hire them all, but I tried to hire at least one from each family."

These poachers-turned-protectors now roamed the marshlands in canoes and motorboats on a mission to protect the animals. Asked how he secured the trust of local field hands and *mariscadores* to work with an outsider, the Spanish biologist Jimenez smiled and confessed he had consumed "rivers of *mate*."

For months on end, Doug and Kris explored the dirt streets of Mercedes, Concepción, San Miguel, and the other communities that ringed the Iberá wetlands—what locals disparagingly called "our soup bowl." Kris—who had been raised in Venezuela for several years as a child—understood Spanish and outback traditions. She gradually fell in love with this rebellious outpost.

Doug felt comfortable in the rural milieu of Colonia Carlos Pellegrini, the small town that served as the center of his initial efforts to win over locals. Years of backpacking left him with an ability to casually wander open-air markets asking for tips about canoe routes into the swamps. Sharing a BBQ or swatting back jokes and sparring in the local lingo, he enjoyed the banter. "Doug had a great affinity for the culture. He *loved* people who wanted to preserve their culture. He really respected people who lived in tough circumstances," said Kris. "They had a culture and they wanted to keep it. That was very, very strong in him."

Doug could sleep in a hammock or on the ground. Locals were abuzz with stories of this eccentric newcomer. He was sometimes seen sleeping in the back of a car. Stories were told that he had swum for an hour in waters teeming with caimans. Then there was his diet. He never smoked and ate little fried or deep-fried food. He rarely drank alcohol. Doug followed a Spartan tradition, *mens sana in corpore sano*—a sound mind in a sound body. Everyone noted his stylish outfits: white beret-style cap, boat shoes, and button-down shirts. Locals found him exotic in a dozen ways, and never boring.

At one aristocratic Argentine wedding ceremony where he was an honored guest, he arrived early and, following his love of profile photography, began snapping portraits of the guests. Assuming the camera-crazed American snapping away was the marriage photographer, wedding guests began ordering him about. Tompkins played along. When it was time to take seats at the main table, there was a missing VIP—where was Mr. Tompkins, the wealthy American? As Doug sheepishly put aside his camera and took his seat, the guests were mortified. The man they'd been ordering about like hired help was in fact their guest of honor. Tompkins grinned, as he loved a good practical joke. Photography was also his way to hide in a crowd. Not one to chit-chat at big social events, he used his camera as a shield and subsequently became quite skilled at shooting portraits.

As Doug barged into ideological and political battles, enjoying the chaos and uproar, Kris hired powerful local women to wage the ground campaign. "Kris is a person who always thinks of the contingencies, but Doug never did," said Heinonen. "That gave them a kind of balance. I think Doug could rest easy a little bit, knowing that he had Kris by his side. Doug without Kris would have been like a rocket headed for the moon. Kris anchored him, and gave him order."

Marisi Lopez worked as a receptionist in the Iberá office until Doug and Kris realized her particular talent: she could read the local landscape like a politician. Lopez became key in building bridges to the local political infrastructure, particularly with the senator Sergio Flinta. As Marisi took over more responsibilities, things took off. "Doug never put limits on the people he hired," she remarked. "He liked to hire strong women and give them a lot of responsibility. He was so sure you could do more than even you imagined. He'd throw you in the deep end of the pool and tell you to swim. If you swam he'd celebrate and if you needed help he would blame himself for throwing you in."

Teresita Iturralde, a lawyer working for Tompkins, rallied locals as she described the synergy of rewilding Iberá's lost species while also cultivating a tourism-based economy. Accompanied by stunning photographs, Iturralde responded to questions and inquiries that erupted in the wake of Tompkins's bold moves.

Tompkins pelted the business elite of Corrientes first with criticism, then with an outright challenge to their dominion over the land. Tompkins loved being the underdog and enjoyed challenging the local captains of industry to a competition on their home court. Rice barons, pine plantation owners, and cattle ranchers in and around the marshlands were accustomed to ignoring bothersome environmental laws. They considered their fiefdoms private affairs, and had no idea what was about to hit them.

Until Tompkins began his impertinent questions and flyovers, few people protested the powerful business leaders, who entered and altered the wetlands as they wished. "Doug was the first to give us a kick in the pants and say, 'Correntinos, look what you've got here,'" said senator

Flinta. "I told Governor Colombi, 'Hey, we've got to do something. We can't have this gringo coming here telling us when we ourselves know what we have to do.'"

After surveying Iberá from the air, Tompkins realized that ranchers had dredged mile after mile of illegal causeways into the wetlands, allowing cattle to walk along the embankments. As the cows grazed, they fouled the water, and the causeways interrupted the natural flow of nutrients and the migration of native species. Massive rice paddies were also being carved out, then boxed in by four walls that further strangled the ecosystem. These activities were clearly illegal. Given the importance of the Paraná River and its greater watershed to the local economy, a strict code of water-usage laws supposedly regulated any alteration of natural waterways. The laws designed to prohibit such projects, however, were rarely enforced.

From his cockpit Tompkins raged at the sight. He calculated that the Iberá wetlands had perhaps ten years left. Without a structured counterattack in defense of nature, a critical mass of habitat would disappear, making any efforts to rewild the landscape futile. Tompkins took action: he hired the best lawyers in town.

Lawsuits in Corrientes? To defend nature? Nobody did that. Land disputes were settled with beer, bullets, or bullying. "I told him that in Argentina, we go to the press, or we launch demonstrations, but we don't believe in the justice system. That could take ten years, and nobody pays any attention; the judges change and they're all corrupt," said Heinonen. "He said, 'I don't care, we're going to sue them all!' I hired three lawyers and we began fifty-four lawsuits, against all the neighbors, and against the state. We set off a storm!"

In court, Tompkins won round after round. In a memo to his staff, he described "the mismanagement and stupidity" of a Señor Maciavelo, the owner of Forestal Andina, who attempted to defy court rulings that his eight-mile-long dike into the wetlands was illegal. When the courts ordered Maciavelo to destroy his dike, Tompkins celebrated. "For the first time in history," he declared, "citizens have risen up against this Far West Do-What-You-Like attitude and taken Sr. Maciavelo to court and won every single legal case. One-two-three-four in a row! The law is clear and Mr. Maciavelo has been given seventy-two hours to begin dismantling the dike and restoring the damage or go to jail. This is a positive sign that the justice system can work in Argentina, despite the millions of complaints we hear from rank-and-file citizens."

Flying continually over the illegal embankments, Tompkins photographed every truck, every work crew, and told anyone close enough to listen that he was going to kick the developers out of this natural paradise. From now on, he declared, the future of Iberá was going to be as a sanctuary for wild animals. Local workers, who had long been disenfranchised and lived on the brink of poverty, pricked up their ears. No Argentine millionaire was going to bat for nature or challenging the elite. So where was this *yanqui* headed?

With the promise of an environmental revolution and the pile of more than fifty lawsuits against the area's business elite, Tompkins was already on shaky ground when a surprise visitor arrived in Iberá—the Chilean president. "Eduardo Frei arrived, talking badly about Doug," said senator Flinta. "Frei came to Corrientes; he met with our governor. I was in the meeting, heard it myself. Frei said Doug would try to take all the water, and that they had cut Chile in two. Frei told people that Doug was a devil."

Chapter 13

Pumalín Park

Doug's enemies really wanted to kill him! They controlled the press. They had all these crazy, insane stories about what he was doing that were just floating around. We would go way up into these fjords on these beautiful hikes. We'd have lunches and he would just crash, in the middle of the day, which was unusual for him. He had the whole weight of the world on his shoulders. They were trying to make him pay. They were doing any dirty tricks that they could possibly do to get this guy out, and he was so tough that they couldn't get him out, but I could tell—they'd seriously zapped him.

—DAN IMHOFF, Doug's son-in-law and fellow environmental activist

By the fall of 1997, Tompkins had more Chilean enemies than he could count. The onslaught, led by President Eduardo Frei, included Catholic bishops, Navy admirals, and a conservative press empire known as *El Mercurio*, which was owned by Augustin Edwards, a powerful ally of General Pinochet. After five years of ongoing battles to build Pumalín Park, Tompkins felt surrounded. Half his waking hours were spent in defensive actions, slowing progress to create the park.

After months of repairs and rebuilding the municipal ferry docks, Tompkins presented the community of Caleta Gonzalo—population:

several hundred sea lions and fewer humans—with a new, properly graded boat ramp. The small ferry crossing included safety lighting, landscaping, and, as a courtesy to local customs, a flagpole.

After the Chilean flag was raised in a humble ceremony, the local mayor—a former political operative with a decade's experience in the brutal El Salvador civil war, a politician named José Miguel Fritis—struck back. Fritis forbade Tompkins from flying the Chilean flag. He accused the gringo of infringing national sovereignty. The battle became personal. Tompkins bought land that Fritis lost at auction. Fritis became the eyes and ears of the federal government in a campaign to watch Doug's every move. When he adopted a new dog, he mockingly named it "Doug."

To better surveil Tompkins, the Chilean Navy built a new office astride the boat ramp. President Frei even approved construction of a police station in Vodudahue, population: less than 100. Tompkins joked that the remote police building should be named in his honor, and instead of fighting the measure he offered to donate the lands needed to house the remote outpost.

"One thing you have to highlight is that Chile has a rule of law, because in Brazil, for something like this, they shoot you. In Peru, they shoot you. He wasn't shot, because this is Chile, but they really hated the guy," said Juan Pablo Orrego, a Chilean environmentalist who worked closely with Tompkins. "Plus, he was intelligent, a millionaire, and a good businessman. They should have respected a guy like that, even admired him. They hated his guts. He was a person like them, but on the other side of the road. He funded the birth of the environmental activism movement in Chile. So this guy was an enemy."

President Frei ratcheted up the pressure against Tompkins. He ordered tax authorities to double down, to scrutinize every paper Tompkins or his organizations had ever signed. "Find something!" the tax officials were told. President Frei's obsession to unearth ulterior motives and discover Doug's sordid secrets ran into a brick wall for the most obvious of reasons: there was no master plan. Yes, Tompkins purchased large chunks of forest in Patagonia, and when he saw a piece of land he liked, he bought it. Sometimes he spent millions fixing it up. Sometimes he sold it later. His top aides knew he was improvising, and capable of more work in a week than most humans could pack into a month.

Despite outward gestures of calm, Tompkins was seething. He hated living under the microscope, especially when he was so certain he was right. He found solace in the sky. In the cockpit of his airplane, Doug soared above Mount San Valentin, Patagonia's highest peak. "He left his mundane thoughts behind and concentrated on flying and the beauty of what he was looking at," recalled a fellow pilot, Rodrigo Noriega. "That mellowed him. He would go for a spin and look around. . . . Part of his masterminding was that he could fly and see things from above. He wouldn't have had that total vision if he had been a flatlander."

When he needed to assess the potential of new lands, Doug circled over them for hours or skimmed the skin of the Earth barely twenty feet above the tree line as he followed hillside contours, the slope of a river, or the depth of a valley. Based on the view from the air, Doug would come up with an architectural masterplan, explained architect Rojas. "Even if the plan wasn't on paper, he would get it clear in his head."

Rojas spent entire afternoons flying with Tompkins as they designed buildings, hiking trails, airstrips, and organic greenhouses for the burgeoning infrastructure their park projects required. "When you're building in these rural, remote places, you don't have neighbors, or streets, or zoning laws, or building codes," said Rojas. "But in these places, you have other laws, which are natural laws: the views, the slopes, forests that could protect you from the weather. Doug figured these out from an aerial perspective."

At Reñihue, Doug and Kris designed an organic greenhouse and built a schoolhouse where teachers educated several dozen students in a classroom heated by wood stove and lit with candles. They even launched a beekeeping project that produced every month thousands of pounds of organic honey.

Realizing that a stable workforce came in the form of a family, Kris and Doug recruited local couples. For families with small children, the attraction was irresistible: quality schooling, purposeful work, and a boss who paid on time. Like his policies at Esprit, Doug didn't offer the highest wages, but the benefits were excellent. Health insurance, retirement funds, and paid vacations were all respected. Workers facing a medical emergency could count on Tompkins to fire up the Husky and fly them through storms to reach a hospital. When a father needed to witness the birth of his child, Tompkins would fly him there.

Tompkins invested hundreds of thousands of dollars into the local economy. He hired metalsmiths and stonemasons to build the administration building and lodge at the entrance to Pumalín Park. Woodworkers carved signage and carpenters crafted furniture. "It was a beautiful process, because it gave dignity to the artisans' work, and when it was finished the people were proud of their work and liked to show it to others," said Francisco Morandé, a young architect working with Doug on the growing collection of farms and park-related buildings.

Tompkins was a stickler for the smallest details. Inside his park signage had to be pleasing and elegant. When painting buildings in Pumalín, he was so fussy about the colors that the local hardware store added a new shade of green to their inventory. It was just easier to set the color, rather than color correcting over and over again until it was just so. They called it "Tompkins Green."

For the placemats at his dining room table he commissioned hand-sewn creations from local weavers. Local metalworkers hammered out the copper-plated exhaust hood for the kitchen. Milk pitchers and flower pots were purchased in Quinchamali, a Chilean village known for pottery. With no electricity on site, local craftsman used hand drills, post-and-beam techniques, plus oxen to accomplish tasks usually powered by electric drills, cranes, and backhoes. They built with little hand axes and handsaws. That's what they knew how to do, and they did it incredibly well. Tompkins was fascinated by their skill; he loved how the artisans could make anything by hand.

In showcasing a touch of local beauty, Tompkins also provided a spark to keep these crafts alive. He even created a series of posters celebrating the backhoe operators as artists. "How do you make the guy who's driving the tractor feel like his work is valuable and adds up to something and is a creative and important act? It's getting that guy to see that the way he drove the tractor was part of this bigger project," said Nadine Lehner, an aide to Doug and Kris Tompkins. "Doug was often working elbow-to-elbow with the people doing very ground-level work on these projects. It made people feel like their work was visible and valuable."

On a scale matched by few foreigners in the region, Doug and Kris invested in frontier culture. Doug donated money to a church radio station, and convinced Monsignor Juan Luis Ysern to broadcast a daily show promoting environmental conservation themes. Behind the scenes, Doug and Kris bought soccer equipment for the local team and an accordion for the town band made up of firefighters. "He went to meet the people, to see how they lived, who they were," said Ingrid Espinoza, a forestry engineer whom Doug had hired. "One time I prepared everything for a visit to a property he wanted to buy. It was a ranch house, very simple, but the owners had prepared a barbecue with baked lamb and potatoes and were all ready to sit down with Douglas and talk business. It was so pleasant; Doug didn't feel uncomfortable. He was never in a hurry, in that sense, despite his being very accelerated normally. He took all the time in the world. He ate the lamb, the potatoes, he drank the *mate*, and I never saw him remove the straw from the *mate*. Never!"

Despite the burgeoning relationship with neighbors and support from Chile's nascent environmental activist movement, Doug's allies in government were few. Audits by tax authorities continued unabated. But he held a trump card: he knew the Chilean government craved to be the first South American country to sign a Free Trade Agreement with the United States. If they pulled it off, billion-dollar investments would arrive. Few questioned that the money would fund environmentally destructive projects including copper mines, timber clear-cuts, and sulphur dioxide–spewing aluminum smelters. The Chileans were begging for a free trade agreement with the United States.

Tompkins had both Prince Charles and Ted Turner on speed dial, and knew how to play politics at that level. At Esprit he'd spent two years negotiating trade treaties with Chinese officials, so he knew the Chileans were not willing to derail a high-profile investor from California. Tompkins saw through their bluff. He was certain that the Chileans were not going to risk the fallout seizing his investments would cause just as they were soliciting billion-dollar commitments from the captains of US financial markets.

After years of public feuding, in 1998, Tompkins signed a cease-fire agreement with the Chilean government. In order to gain peace, Tompkins promised to avoid buying any more lands for twelve months. "The agreement we signed is a protocol, not a binding document. It makes the government comfortable, and helps push this project along," Doug wrote in a letter to staff. "It's a step in the right direction. I'd say another eight years of serious work lie ahead. We used to spend half our time fending off attacks of all kinds—tricks, threats, and criticism. Now, we can concentrate ninety-eight percent of our time working on the park. We achieved a little blow for the environment. A drop in the bucket. But it's going to take lots of drops in that bucket from a lot of people to become a corrective force looking after our world."

Hardline advisers to President Frei sought to muzzle Tompkins and wanted the cease-fire to include prohibitions on Tompkins's taking any actions or making statements against the US-Chilean Free Trade Agreement—and to avoid criticizing the president. The attempts to muzzle the renegade American via legal accord tanked. Working with Chilean activists Carlos Cuevas and Patricio Rodrigo, Tompkins was learning to build bridges into the government bureaucracy, particularly with Ricardo Lagos, the public works minister and a man on the short list to become the next president of Chile. Cuevas and Rodrigo were passionate activists and each had worked inside the public lands ministry. Tompkins valued their insider POV and considered them key allies.

Powerful forces at the US embassy, including Ambassador John O'Leary, publicly hosted Tompkins at the embassy and made a point of flying south to attend the annual folk festival sponsored by Kris and Doug. Along with some 300 guests who attended the festive, buffet-style event, Ambassador O'Leary wandered the tidy pastures, feasted on homemade apple pie, and admired the stunning attention to detail throughout the parklands. Whether it was a small bridge or a creek or the black-and-white photographs decorating the lodge, the entire project radiated beauty.

Doug was advancing on his park plans and now had seven families living full time on his ranch. "We had thirty projects going and we worked together all day, dined together at night," said Morandé. "We designed everything from the lodges and restaurants to the doghouse."

Although he was half a world away, Tompkins kept close tabs on his foundation back in San Francisco. He remained committed to funding a worldwide network of environmental activists. His grantmaking, run through the Foundation for Deep Ecology, continued to donate $3 million a year. Tompkins loved to fund the underdogs. Many of the grants went to small, tightly focused groups like those that sought to block road building inside national forests and those that fought to protect beluga whales in the Arctic. These small startup funds allowed dozens of environmental activist communities to build their organizations, pay the rent, organize seminars, and publish books. He also funded million-dollar newspaper ad campaigns, including an uprising he stoked against the World Trade Organization.

Throughout the late 1990s, as he built his pioneer village at the tip of South America, Tompkins commuted regularly to San Francisco. He chaired board meetings at his foundation, visited friends, and helped strengthen the International Forum on Globalization, a think tank run by his friend Jerry Mander. Tompkins fought for citizens and the environment as he sought to unmask what he saw as the hidden hand of global capitalism. Specifically, he sought to warn of the rising power of global corporations. Together with a team from the Public Media Center and Andy Kimbrell from the International Forum on Globalization office in San Francisco, they designed a series of full-page advertisements for what they called "The Turning Point Project."

The $60,000 full-page advertisements ran in the *New York Times* and *The Wall Street Journal* and challenged the power of corporations that avoided taxes or sought out the weakest link in environmental protection agreements. These corporations—often publicly traded on the New York Stock Exchange—held loyalty to no nation and behaved as if the planet was their oyster. Tompkins and Mander put a name on the enemy: The Invisible Government.

In 1999, the World Trade Organization's annual gathering was scheduled for early December in Seattle. In the weeks leading up to the WTO meetings, Tompkins ran a series of Turning Point ads in San Francisco, Portland, and Seattle newspapers. He bankrolled the IFG's massive teach-ins that gathered some 3,000 antiglobalization activists each day for strategy sessions. Tompkins put up the cash to rent Benaroya Hall, home to the Seattle Symphony and a building with the finest acoustics in town. He paid for everything from guests' airfare to the rented microphones. Day after day, this critical mass of impassioned activists became ever more fired up as they shared stories. "All these activists from around the world were involved," recalled Mander. "It was a gigantic, noisy success. It got press attention, and then people poured onto the streets after that. Seattle got really turned on against the WTO."

When the WTO meeting began on November 30, activists shut down the city of Seattle with civil disobedience in the spirit of marches lead by Mahatma Gandhi and Martin Luther King Jr. Tens of thousands of people—labor activists, students, and environmentalists—took over the streets and trapped WTO delegates in their hotel rooms. The Seattle WTO protests were raucous, largely nonviolent, and hugely effective. The activist group cofounded by Earth First! leader Mike Roselle and dubbed The Ruckus Society had achieved its goal of "a loud, angry interruption, a hullabaloo, a disruption."

Images of American youth and workers rejecting the dictates of the WTO flew across newswires worldwide. Without a public word, Tompkins had planted one of the seeds for the uprising. "It was amazing," reflected Mander of the Public Media Center. "We brought 100,000 people out onto the streets of Seattle, and shut down the WTO meeting. That was a fantastic event."

Back in Patagonia, Doug and Kris were gaining international attention for their conservation plans that now branched into the newly flourishing field known as rewilding. As a conservation strategy, rewilding allows ecosystems to reassemble key flora and fauna, and begin recovery. As practiced by the Tompkins team, it also included the removal of non-native species. Following a clear-cut, there was no recipe for seeding the diversity and equilibrium that a mature forest might finally settle upon after thousands of years of evolution. Despite the government harassment, Doug and Kris now owned nearly a million acres in Chile and Argentina, and they had a huge opportunity to try out their rewilding strategies on the dozens of forests and farms they had now purchased.

Doug put his money into the idea that environmental restoration depends on a rethinking of modern agriculture. Monoculture plantings, intensive insecticide, ignorance of overall soil health, and a growing world appetite for meat made it apparent to Doug that without a farming upheaval, his plans for the wild would be doomed. In Argentina he purchased run-down ranches and began to design organic farming at a 18,000-acre riverfront farm known as Laguna Blanca. Day after day, Doug swirled in his plane above Laguna Blanca, like a painter obsessed with a blank canvas; he sketched his visions on a scale that only a pilot could imagine.

By 2000, Pumalín Park was the largest private park initiative in the world, and supported by a small group of donors who traveled to Patagonia to study the land and meet with this unique couple living off the grid. Arriving in southern Chile, the donors found that Kris and Doug were exquisite hosts. Doug lobbed controversial intellectual arguments, bandied one-liners, eviscerated mainstream attitudes, and often offended an entire room without stopping to take a sip from his decidedly nonalcoholic glass of warm water (no lemon, no tea). Kris was far more than peacekeeper. She was the secret administrator of the projects, building bridges with influential donors, smoothing the ruffled feathers, and sorting through her husband's wild ideas.

In the morning, he says, "Come on. I want to show you what the locals call the land of a thousand waterfalls." And he's got a couple of planes. One of them was called a Husky. It had a canvas skin. And it's just two seats, one right behind the pilot. It's not an ultralight plane, but it's small—And so we go up and we're flying in these narrow canyons and looking at these waterfalls and Andean condors. And we didn't feel much bigger than the condors up there. After he showed me all the beautiful waterfalls, we fly back seaward, and he says, "There's a hot spring down there." And he buzzed over it and I can see the piles of rocks people piled up, to have a hot spring pool to lay in. I see the tiniest little sandy beach. And I realized, Oh, Christ. He's going to land on that. *And I'm thinking,* He probably could land. I mean it's really short, but I suspect he's done this before. *He drops down and he catches the edge of the sand and lands. It must have been fairly packed sand. And we go over and strip off our clothes and jump in the hot springs for about twenty minutes. We're talking about saving the planet and all of our campaigns. Then we get out and dry off on the rocks for another twenty minutes and throw our clothes back on and jump back in the plane. He taxies us back to one end of the runway such that the back wheels were in the water. And he's gunning that engine to the point I'm thinking it's going to blow up. Then he pops the clutch, so to speak, and we take off safely. Later that night, he says, "I don't know if it's fully legal to land in a place like that but, it was so beautiful, I just needed to take a hot bath.*

—RANDY HAYES, founder,
Rainforest Action Network

Doug and Kris now had a solid infrastructure on their parklands including cabins, a lodge, a visitor center, and a bevy of locals working to guide hiking expeditions and maintain the park grounds. Doug spent months walking the land to understand where to situate trailheads and where to place the campgrounds. He sculpted the hiking paths with the same attention he gave to window displays at The North Face or the curve of a font in the Esprit catalog. The few climbers and backpackers who ventured to Pumalín Park discovered a world-class natural sanctuary in its infancy. Doug like to call one hidden corner of his valley "a hidden Yosemite."

The election of socialist Ricardo Lagos to the Chilean presidency in 2000 sent Doug and Kris's possibilities of completing Pumalín Park soaring. Lagos was an ally beginning a six-year presidential term. The left-leaning politician had been forced into exile during the military dictatorship and worked abroad, teaching political science at the University of North Carolina, Chapel Hill. He was a courageous rebel who, like Tompkins, showed unusual poise while navigating extreme risk. As a young socialist leader, Lagos was placed high on General Pinochet's hit list. Following a failed 1986 assassination attempt on the dictator near his summer estate, secret police death squads went hunting for Lagos. When Chilean detectives got word of his whereabouts, they arrested him and, in order to protect him, placed him under heavy guard in a jail. The detectives were hoping to keep Lagos out of the feared torture chambers that many Chileans called "concentration camps."

After Chile's return to democracy, Lagos become the minister of public works, and jump-started the Chilean economy via a multibillion-dollar highway and bridge program. Given Chile's dependence on the export of raw materials, including cellulose, copper, fishmeal, and fresh fruit, Lagos invested his ministry's resources in infrastructure ranging from national cell phone coverage, modern airports, multilane highways, and container-friendly ports. Tourism, he recognized, was also going to become an important part of his country's development. What other country could offer visits to geysers in the world's driest desert or support cruise ships en route to Antarctica with stops at wineries and Easter Island along the way?

President Lagos provided Pumalín with legal protection, not as a park but a "nature sanctuary." Despite intense opposition, bureaucratic sabotage, and unusual delays, Pumalín Park gained its first degree of protected status. Under Lagos, doors slowly opened for Doug and Kris at government ministries that had previously ignored them. Doug ceded to the requests of his lawyer and provided private briefings to Chamber of Commerce executives, tourism officials, and reactionary elements within the Chilean government.

At a VIP meeting at Chile's presidential palace, Tompkins was scheduled to meet the head of the navy, and etiquette dictated he wear a tie. "Doug was really struggling with that because he was a rebel at heart," laughed Daniel Gonzalez, his aide. "He kept saying, 'Why on Earth do I have to wear a tie to visit with these people? I don't wear ties!' I remember him putting a tie on at the very last minute before going in, and as soon as we left taking that tie off."

"Doug made a presentation about his ideas to the navy—his project of creating national parks," recalled Pedro Pablo Gutierrez. "Afterward one of the officers apologized, saying, 'Look, all that nonsense we heard about you? Now that we had the chance to meet you, we realize we were taken for a ride.'"

Tompkins spent much of his waking hours working on a book documenting the destruction of Chilean forests. Entitled *The Tragedy of the Chilean Forest*, it was essentially a Chilean version of his *Clearcut* book created years earlier. The project was the result of hundreds of hours flying above Patagonia in small airplanes. From his cockpit, Tompkins snapped thousands of photographs, then in the evening studied the remaining forests. One evening, looking at the pictures, he discovered a rare stand of old-growth coastal cypress—known as Guaiteca—in a bay called Bahia Tic Toc. With no access by land and rough seas off the coast, much of the area had never been logged, and what had been destroyed hadn't been touched in decades; Tompkins saw that natural regeneration in these forests was well on its way. He scouted about—who owned Tic Toc?

Flying over the region again with his friend Peter Buckley, Tompkins pointed out tracts of land for sale. The vast wilderness offered majestic possibilities for conservation projects. Tompkins was talking a mile a minute and so enthused that Buckley joked it was like flying with "an overcaffeinated real estate agent." When they zoomed over Corcovado, a volcano-shaped peak with the crystalline waters of Bahia Tic Toc below, Buckley was mesmerized. "The moment I saw this big rock, kind of like a miniature Sugar Loaf in Rio de Janeiro, and this beautiful bay and granite islands with a river coming down, I said, 'Well, if this is available, I'll buy it.'"

One week later, Tompkins informed Buckley that the volcano, the bay, and the surrounding forest were all for sale. The complicated part,

Tompkins informed his friend, was the money transfer. They needed to buy a company in Lichtenstein, then buy a Panamanian company that owned one asset—the land in Chile. "That's how Pinochet cronies covered up their ownership and interests in these things," explained Buckley. The entire deal was consummated in record time, and Tompkins suddenly had a critical piece of land for building Corcovado National Park.

"When Peter Buckley agreed on plunking $1.75 million for a big chunk of wilderness and forests down here, it was as true a pure gesture as I've ever seen anyone make," Tompkins wrote his friends. "He absolutely spontaneously said yes when I asked him. I was very touched, I can tell you. That was a bright spot in a bit of dismal atmosphere!"

After further studying the maps, Tompkins realized that much of the land surrounding Buckley's purchase was owned by the Chilean Army and Navy. He'd had an inkling as his caretaker described submarines, Marine invasions, and battlefield practice that left the pristine beaches littered with copper shell casings.

Conservative elements of the Chilean government described Bahia Tic Toc as "a secret submarine base," but Tompkins knew better. It was a remote outpost with zero strategic significance and a rich biodiversity. He focused his efforts on building a new national park. If he couldn't overcome the hurdles at Pumalín Park, perhaps he could break the logjam with Corcovado.

I got to know him well, and saw that his breakfast consisted of hot water. I said, "Aren't you going to put some lemon in it?" He replied, "No, just hot water." And I thought, This guy is very weird. Hot water for breakfast? *He was a man with exquisite taste, yet who dressed in ripped blue jeans. How do you pull those threads together? He ordered sophisticated furniture made in Buenos Aires and he was an aesthete to the max, but not with himself. He walked in with shoes so worn, let me tell you. I know that the more shoes are worn, the more comfortable they are, but I could never go around with those shoes. How do you reconcile it? He was a man of good taste, refined, who read a lot, was up to date with everything. He always bombarded me with the things he was reading, and I'd have to read it or he would think I was an idiot. And I am not accustomed to being taken for an idiot.*

—RICARDO LAGOS,
president of Chile, 2000–2006

"We maintained a good relationship when I became Chile's president," said Lagos. "The tipping point was one day he came to visit me and he said, 'Mr. President, I've come to propose some business.' 'Great,' I replied. 'What's the business?' And he replied, 'I've got 80,000 hectares in Corcovado, the Army has 80,000, and you have 100,000 hectares' ['you' meaning the Chilean state]. 'What do you think of us creating a park? I'll give my 80,000, you convince the Army to give theirs, and you give the other 100,000.' I replied, 'Let's do it!'"

President Lagos instinctively trusted Tompkins, and was wary of the military, given the way it had tortured and even executed some of his political colleagues in the early years of the Pinochet dictatorship. Although he never called the creation of Corcovado National Park an act of revenge, President Lagos knew that every Chilean president since the 1920s had created a national park, and he himself relished the idea of converting military land into a national park.

The joint collaboration was completed without any of the battles and hassles that shrouded nearly every move at Pumalín Park. Virtually no settlers lived in the zone, the land titles were clean, and the sheer beauty of the snowcapped Corcovado volcano made it difficult to resist. Lagos prodded and pushed. He knew exactly how to manipulate the levers of power, and he made a point of inviting Doug to sit next to him at the celebratory dinner in Puerto Montt when the park was a done deal. Tompkins, for his part, was agog. In a letter to Buckley, he described the scene:

> I went for a long walk on the beach with the president and had some good conversations about many subjects but primarily on environmental protection. I began the conversations about a second national park that would be ready in his second term! He laughed about this and said, "Okay!" Maybe all this work is finally paying off and the insults to my integrity, the dirty political maneuvers, the low-class moves on the part of small-minded bureaucrats, the carping details, the agonizing delays, the huge costs, my temper getting short even with my darling wife—all could be worth it.

After Corcovado National Park was dedicated, Tompkins sent Buckley a stack of pictures highlighting President Lagos, the military ships, and the park inauguration ceremony. But he didn't realize he had never invited Buckley. "I said, 'Thanks a lot for inviting me,'" Buckley recalled. "He said, 'I'm sorry, I guess I kinda screwed up.' I said, 'You're not off the hook. I want a thank-you note from Richard Lagos.' He said, 'I can't do that.' I said, 'Oh, yes, yeah you can. I'm serious, I want a note from Richard Lagos. You can call him.'" Months later Chile had its brand-new Corcovado National Park, and Buckley had his signed thank-you note from the president of Chile.

Chapter 14

In the Heart of Patagonia

Where we live in South America, we're still on the agricultural frontier. They're still taking down forests and converting to ploughed agriculture. You see it happening very fast. What's happening in those habitats? They're disappearing! Where is the space for the other creatures that are supposed to share the planet with us? Why do we have a whale sanctuary in the far South? We are trying to hold off the human project, to keep it at bay.

—DOUG TOMPKINS

Like the stray dogs so common in Chile, the Valley Chacabuco ranch was scruffy, filthy, and beaten down when Doug and Kris first visited. More than 25,000 sheep ranged the valley, eating all grass, shrubs, and even seedlings. Everything green was gone. The valley's flowing prairie grass, the native calafate berries, winter's bark, and wild strawberries had been eliminated. Herds of sheep—"hooved locusts," in the words of the naturalist Edward Abbey—were fenced in by hundreds of miles of barbed wire, and the ranch employed a team of hunters, known as "lion men," who shot, poisoned, and trapped wild puma.

In the high-mountain lakes, flocks of flamingos bathed, and at dawn, buss-necked ibis with their striking curved beaks searched for food in the grass and screeched a warning to the other animals whenever humans arrived. Tiny Austral pygmy owls sat on fence posts. Thousands of feet high in the sky, Andean condors rode the updrafts created when winds off the Patagonia ice field collided with the flanks of mountains. Patches of beech forest, small packs of wild South American camelids known as guanacos, and the string of twenty high-mountain lakes buzzing with birdlife made it possible to imagine its earlier splendor, yet by 2004 "ValChac"—as the ranch was known—was hemorrhaging both money and biodiversity. "When I drove through the Chacabuco Valley for the first time, I saw the extra-high 'guanaco fences' designed to keep these first-rate jumpers out of the best grasslands, which were reserved for the cattle," said Kris Tompkins. "My eyes glazed over, looking out on the tens of thousands of sheep grazing the bunch grasses up and down the valley. The grasses looked patchy and dead. Nothing left for wildlife."

Kris imagined a grand plan to revitalize ValChac's degraded lands. She seemed smitten by the challenge, and together with Doug they had ogled the ranch from the air, taking so many high-resolution aerial photos that when a ranch owner named Francisco De Smet accepted an invitation to their Reñihue homestead, he discovered that the couple had thoroughly mapped his ranch with their aerial photos. Like secret admirers from above, Doug and Kris imagined life with the Valley Chacabuco ranch under their care. Could they promote native flora and rewild the fauna? Yvon and Malinda Chouinard shared their passion, donating land and money, and threw their considerable influence behind the park project.

After years of wrangling, the sale was formalized in October 2004. De Smet wanted $9 million for the land and $1 million for the animals. Kris fought hard to buy the ranch without the 25,000 sheep, but to no avail, as De Smet didn't want the animals either. Both parties knew that dumping that many sheep into the market would collapse the price, harming other sheep farmers, and there wasn't nearly the capacity for the refrigeration or the trucking for such volume.

After Kris bought the ValChac ranch, rumors ricocheted through the valley—"the environmentalists are coming to shut the place down!" Locals had heard and read about the battles in Pumalín, a few hundred miles to the north. The same gringo couple "who cut Chile in two" were taking over the crown jewel of ranching in the very heart of Patagonia. Would they eliminate an entire lifestyle?

Their fears were not unfounded. Colonia Dignidad, the same German enclave that Doug had driven by during his epic 1968 road trip, then again with the colleagues from Esprit to raft the Biobío, was now being exposed as a clandestine torture center utilized by the Chilean military. On the surface Colonia Dignidad had promised free education and medical care; in reality it was a house of horrors run by a cadre of reclusive Germans. For many powerful politicians in Chile, Colonia Dignidad felt like a nation inside their nation. Protected by the German government, it seemed to be an appendage of the Cold War, and the Chilean government was unable to control these powerful foreigners or dismantle their strange customs and violent traditions. More than a few Chileans now suspected that Tompkins might also be hiding a secret agenda.

As she prepared to take possession of ValChac, Kris built a transition team. She gathered a three-person, all-female crew to set up the rules. "We go down to ValChac and we have to formally accept the estancia from the owner, and meet with all the employees, twenty-six gauchos, and two cooks," said Kris with a laugh. "And these are gun-slinging, knife-in-their-belt gauchos. And here we are! Carolina has prepped her employee handbook. And it said, 'No weapons allowed,' and as she reads it aloud everybody looks down at their feet because they're all packing pistols, rifles, and have knives down their pants. And then she said, 'No pets!' And they had seven to twelve dogs apiece. It was a funny way to get started."

Just months after they bought the ranch, Kris and Doug faced a crisis. Thousands of lambs were about to be born. They had to nurse the ill, euthanize the dying, and vaccinate all the bleating lambs. Now they understand why De Smet had fought so hard to leave the sheep behind.

Kris took charge. Before she was Kris Tompkins she was Kristine McDivitt, a descendant of Scottish sheep ranchers. As Doug poured millions of dollars into the 18,000-acre organic farm in Argentina known as Laguna Blanca and spent his days engrossed in the biochemistry of organic agriculture, Kris organized studies of the fauna of ValChac. Malinda and Yvon Chouinard were also deeply moved by the rewilding and restoration opportunities in Valley Chacabuco. They held a deep loyalty to Kris and knew firsthand how valuable she had been at Patagonia Inc. Investing their time and resources, they played a key role in purchasing adjacent lands. As they collectively studied the fauna, debated the best conservation techniques, and bought land, a series of questions arose. Which animals had been hunted to extinction? Did surviving ecosystems for these same animals even exist? Could she reintroduce the endemic species to the ravaged land? The valley served as a biological corridor for species migrating through the heart of Patagonia, making the land an ideal project for her and Doug's growing interest in rewilding.

After purchasing the ranch, Kris led the negotiations. She and Doug offered the lands to the Chilean government for free. The Chilean National Parks service could administer the land, similar to the process in Argentina with Monte León. The Chileans rejected the offering. They wouldn't take a free national park. Daunted by the upkeep and infrastructure needed, they feared the Chilean government couldn't afford to accept the donation.

Doug and Kris turned the surprise rejection into a challenge. Could they restore an overgrazed ranch and in the process rebuild an ecosystem? "The project had a certain romanticism, because starting a project from zero like that meant a lot of field work for Doug and Kris," explained Ingrid Espinoza, who worked with them both and lived, along with her husband, at Reñihue. "We wanted to turn the land back to its natural state, but at the start we had a problem because we had thousands of sheep to manage."

Initially, the only habitable buildings in ValChac were a run-down sheepfold and a shack that had once been used by the Chilean herdsmen. Doug and his team of builders remodeled the sheepfold into a basic shelter, and soon Kris and Doug were sleeping there several days a week as they sketched out the infrastructure for their new national park project.

At dusk, sunbeams sliced through purple clouds as the sky scrolled through a dozen blues. Moonlit skies allowed locals to navigate on horseback. Nights were a jumble of wild animal cries. The guanacos gave a piercing whinny when threatened. "Then there was the chaos of the chase and the screams by whatever animal got eaten," said Kris. Despite frequent puma attacks on livestock, Doug and Kris preferred to sleep in a tent by a lake. Phone connections were nonexistent, so they used CB radio and gave each other call names. He was *Aguila* (meaning Eagle) or *Lolo* (which was slang for young guy). She was *Picaflor* (meaning hummingbird).

Kris felt more alive than ever. "When I go out early in the morning, it's still not quite light and I start to get paranoid," said Kris, describing her dawn walks inside Chacabuco Valley. "But it feels good to be aware. I love that. You're not top predator. . . . I was in the grasslands, which are very up and down, and I thought, *Wow, this is how the ancients felt*."

Now that her foundation Conservación Patagónica owned the vast valley, it was time to explore. What exactly had she purchased? Leaving the house, Doug and Kris liked to walk hand in hand: no phones, no digital devices, and, weather permitting, no clothes.

Doug no longer dropped acid or used hallucinogens. On these naked trips, he needed no psychedelic shortcut to appreciate the beauty of the Patagonian landscape. "Get the humans out of the way," he repeated, "and nature will flourish." Kris and Doug often stopped on their walks to pinpoint the orbit of condors circling above a dead or rotting carcass. When they searched for the object of the condors' attention, at times they found an unnatural sight: dozens of mangled lambs. Mother pumas teach their offspring to bite the jugular, killing the prey at once. But the puma cubs were less efficient, often maiming or wounding the lambs, without killing them right away. The toll from a single evening could reach twenty or thirty lambs, as the mother puma's hunting class left behind a bloody mess. Condors glided overhead, waiting their turn at the carcasses.

When Doug needed to escape South America for wild adventures abroad, he took up an offer from his friend Mike Fay, who needed an experienced pilot to help with an aerial survey he was organizing in Chad, in central Africa. The assignment was about as boring as flying could be—fly straight lines across the desert, estimating animal populations throughout 3,000 square kilometers. The flight path of the survey required hundreds of flights in each direction, like a huge piece of graph paper. Tompkins flew line of sight, deliberately ignoring the instruments. He rarely drifted more than ten feet off course. "It was phenomenal," said Fay. "You can always tell a good pilot from a bad pilot by how straight their lines are—and his lines were perfectly straight.

"He enjoyed the challenge of being able to do that and obviously flying around in wild places and participating in the conservation actions," said Fay. "But at the same time, he found Africa very irksome because he couldn't stand the disorder, the chaos, and his meticulous mind didn't gibe with the African scene."

While circling above Chad with Tompkins piloting, Fay spotted poachers gathered in a camp, armed and on the hunt for elephant tusks. Fay asked Tompkins to dive-bomb the camp, then pull up at the last moment. The maneuver would allow Fay to photograph individual poachers. Tompkins asked if they would get shot and Fay assured him that poachers "always miss with an AK."

Tompkins circled the camp four times at low altitude while Fay snapped away. "I can see the guy pointing his gun at us. And I can see his shoulder jerking and jerking, and I said, 'Hey, Doug, he's shooting at us!' He just responded, 'Oh, wow, really?' He kept flying around. Danger didn't even come into his brain. And we're only two or three hundred feet from this guy. We're not up at a thousand feet. Kris was all upset. She said, 'You can't do that. You can't! You're going to get him killed!' And I said, 'I'm not going to get him killed.'"

Doug and Kris now ping-ponged from their homes in the sodden Pumalín rain forest and the Iberá wetlands to their new project on the prairies of Patagonia. Sloping above the aquamarine Baker River and nestled in a picturesque valley some fifty miles inland from the Pacific Ocean, ValChac offered a sunny, blue-sky alternative to the gray and somber weather in Pumalín.

Every time he piloted his Cessna south, Doug flew a different route as he gathered 7,000 hours of flight time, often without the use of instruments. Flying by sight, Doug lectured his bedazzled passengers, was essential in order to avoid mishaps should the instruments fail, or as happened more than a few times to the overly enthusiastic pilot, when his plane ran out of gasoline in flight.

Among the most threatened by the arrival of environmental activists were the puma hunters, the men who embodied the noble pioneer at the heart of Patagonia cowboy culture. Known as *leoneros* or "the lion men," they were brave hunters who mounted horses on expeditions high into the Andes Mountains, sometimes even into neighboring Argentina. These hunters carried knives, pistols, and shotguns. Their most effective weapon was a pack of dogs, a motley crew of assorted colors, sizes, and species that looked assembled from a dozen random street corners. More clever than purebreds, these streetwise dogs were not only hunters. For the nomadic gaucho who often spent months in the field, they were company.

Doug became obsessed with protecting the pumas. Allowing the pumas free range to hunt and kill was a key part of his strategy to allow the area's apex predator to regain its role, regulating the populations of native wild animals. The Tompkins team re-trained the ranch's lion hunters as park rangers. Instead of killing pumas, they shot them with tranquilizer darts, then hoisted the 100-pound animal atop a gurney while the veterinary team affixed a radio-signal tracking collar. Given the shoddy reception for many of the radio collars inside the creviced valley, the hunters were asked to do what they knew best—track pumas. The former hunters led the field biologists on day-long treks to identify key areas of puma habitat. Forced from the valley floor by a century of bullets, the puma population was scattered and rarely spotted. No one had any idea how many of these mountain lions lived in ValChac.

Doug assembled a multidisciplinary conservation team to study the habitat, health, and remaining population of pumas in Chacabuco Valley. He convinced a Santiago zoo to collect the puma piss from their caged felines. Hauling barrels of puma urine to remote Patagonia, Doug realized, was the first step to helping native mountain lion populations recover. Placing the puma urine in a spray bottle, Doug had his team mark territory with the scent of an unknown puma. He figured that encroaching pumas would roam elsewhere and avoid the easy to capture sheep in the valley floor.

With Kris at the helm of the Chacabuco Valley project, the couple split their time between the stalled Pumalín Park further north and their new operation that they branded as Patagonia National Park. Work usually began at dawn and continued well past dusk. Doug often worked twelve-hour days. Kris felt as if she was CEO again, but this time instead of running the Patagonia company, she was rebuilding an ecosystem in Patagonia itself. Doug stashed love notes in her pockets, shoes, and drawers of clothes. He even taped notes to the back of his pilot's seat. When she sat down in the plane, the note was at eye level. "I mean, it was like Easter eggs," Kris said with a laugh. "Having that kind of love changes every cell in your body. Your face changes. The way you answer the phone. Everything relates to that other person. And we were a little obsessed with one another. Our happy times were just the two of us."

At night by candlelight, Doug studied maps of the region, known as Aysén. Placing photographs side by side by side, he stitched together landscapes. He took these photos while flying with one hand and photographing with the other, often just fifty feet above the land. When he needed to focus the camera, he steered with his knees. Quilting the individual images together, Tompkins was able to peer into hidden pockets of the valley. The Aysén region was poorly mapped, and much of what was known came from the 1912 journey of a mountain-climbing Salesian priest named Alberto De Agostini who spent a decade hauling a 6 × 6 medium-format camera into the mountains. Like Ansel Adams, De Agostini's photographs stirred a broader conservation movement.

Impressed by the discoveries of De Agostini, Salesian leaders relieved him of priestly duties as they allowed him to wander, wonder, and photograph. He was their nomadic emissary into the wilds. Doug understood De Agostini's wanderlust and pored over his accounts. He read the brazen tales of explorers who were so often racked by storms off the coast that they named the area "The Gulf of Pain."

In Patagonia, the wind made travel difficult. In the air, the gusts sent his small plane bouncing through the clouds, and on the road, the wind shoved the car from one lane to another. On the lakes, the wild winds were particularly dangerous. Given the maze of inlets, bays, rivers, and mountains, small ferries were still used to portage cars and trucks, sometimes just two or three at a time. Water temperature in the region was often barely above freezing, and six-foot waves made even these short ferry trips dangerous. When early pioneers found a safe harbor on Lake General Carrera, they named it "Puerto Tranquilo"—the Port of Tranquility.

But it was the blank areas on the map that seized Doug's curiosity. What was out there? Could he discover areas that no human had ever walked, seen, or smelled? As he explored, Doug kept his favorite areas secret—after he spotted them from the air, he quietly purchased the lands if he could. If there were no roads, no access, all the better. Nature, he insisted, needed insulation from global capital and a chance to evolve, unencumbered by the short-term demands of modern industrial society. He followed closely the work of Harvard biologist E. O. Wilson, who was proclaiming the need to save half the planet. In order to ensure a base for species to survive and reproduce, Wilson suggested it might be necessary to set aside a full 50 percent of the planet for nature reserves or refuges. Hearing the plan for "Half Earth" conservation, Doug chuckled and called it "a good start."

If you fly like Doug and myself, most of the time you're only a couple of hundred feet off the ground. You're almost close enough to touch it. And if you want, you can go five feet off the ground. So if you really want to look at something you can go down to that level. It's this three-dimensional space, and you can explore and get any perspective, any scale that you want, pretty much by flying high or low. You can go up every river valley, you can look at every forest, you can see the wildlife population.

Doug had been flying for almost fifty years and he had thousands of hours, so the perceived danger when you're a pilot like that seems very small, and so you can do spectacular things in an airplane that no one else would do. And that's the way Doug was. He was basically like a bird. He is part of the airplane, or the airplane is part of him. And he is constantly moving his body with the plane because it's just one unit. It was an extension of his body rather than his airplane. It's like you put it on and you start flying, so it gives you this incredible freedom that I think birds feel, too. It totally amplifies your humanity because you can do things that a very, very small percentage of people do. It makes you almost superhuman.

—J. MICHAEL FAY, ecologist,
defender of wildlife, explorer

As they mapped the Chacabuco Valley, Doug and Kris understood how it was formed in part by a glacier some 10,000 years earlier, which ground up the bedrock and scooped out indentations that later filled with water. Dozens of lakes and ponds speckled the valley and formed wetlands that attracted pink flamingos, black-necked swans, and bronze-winged ducks. Gray and red fox thrived off the abundant supply of rabbits, birds, and armadillo.

Centuries before the Spanish conquistadores arrived, a native culture known as Oanikenk roamed the valley hunting the guanaco. The Oanikenk diet was rich in protein and calories, and at a time when the typical Spanish conquistador was 5'4" the Oanikenk averaged 5'11". Stunned by the stature and muscle mass of these native peoples, the explorer Magellan called them the "Patagones," which in Spanish meant "Big Feet," and thus the region was christened Patagonia—Land of the Big Feet.

After Spanish colonization, the entire ecosystem was upended by sheep. A 1908 concession from the Chilean government and the dreams of an English explorer, Lucas Bridges, led enterprise after enterprise to seek its fortune in Patagonia. The dream was to turn the flowing grasslands of Valley Chacabuco into healthy herds of sheep that would, in theory, produce hordes of cash. It never worked. The fragile ecosystem had evolved in balance with the grazing and migrations of guanaco, which lived in herds of thirty to sixty members. When tens of thousands of sheep began grazing, the grasslands disappeared and were replaced by rock-studded scrublands where few plants reached knee high. Year after year the operation was unable to make a profit. The riches of the land were ransacked as sheep gnawed away. The lack of roots and vegetation left topsoil exposed, and after heavy rains, the increased erosion led to the eventual death of streams and rivers. Fish populations plummeted as the silty waters diminished oxygen supplies.

The Chilean government further subsidized the destruction of the region's hardwood forests. Settlers were offered title to the land if they cleared away the woods. As German, English, and Chilean ranching operations moved into the area in search of an economic foothold, the Chilean government required that 50 percent of the lands be cleared for animal grazing or crops. The easiest way to eliminate the hardwoods was fire. "They said to the colonists, 'You want 300 hectares of land? You have to burn 150 hectares.' And then the winds would push the fires. They burnt seven million acres of forest in Aysén," said the environmentalist Juan Pablo Orrego. "There were fires burning for over ten years. It even melted the permafrost."

Prioritizing their restoration tasks, Kris zeroed in on the fencing as her first obstacle. The fencing blocked the migration of guanacos. Able to leap eight feet high, the animals attempted to cross the valley despite multiple strips of barbed wire. Hundreds bled to death, impaled on the spikes. As long as the fencing existed, it was impossible for wild animals to reestablish natural migration patterns.

Kris wanted the guanacos to come down from the hills, the puma to stalk and kill them, and the wild ostrich–like Darwin's rhea to bound across the grasslands eating frogs, mice, and plants. To do that she needed to remove 400 miles of fence, so she launched a volunteer program. Working with Malinda and Yvon Chouinard, she coordinated so that employees from the Patagonia company were given time off to volunteer. With thick gloves to protect their hands, they dug up fence posts and coiled the barbed wire in large balls that looked like galvanized tumbleweeds.

The volunteers also waged war on invasive species. Thousands of plants were dug from the soil—roots and all—and then carted away or burned. "It was so interesting to see how the local people reacted to fence removal," said Lito Tejada-Flores, Doug's longtime climbing friend who had bought a plot of land nearby. "I don't think they'd ever seen that much space without fences. There were campesinos coming from other areas just to see it."

When Kris and Doug's friends came to visit the Chacabuco Valley, some fell in love with the region and bought property nearby. Edgar and Elizabeth Boyles, Doug's close friends from San Francisco, purchased a place in Puerto Guadal, where their son, Weston, could explore rivers and wilderness and accompany them to lively dinners at the Tompkins home. From the time he was ten years old, Weston heard the same lecture from "Uncle Doug"—skip college and dive directly into life.

Doug's grandson Gardner Imhoff spent his vacations in ValChac where he hiked, rafted, and listened as his grandfather attempt to convince him that "Harvard is a rip-off and ought to be boarded up with cheap plywood." Never a fan of formal education, Doug delighted in upending the carefully laid plans of parents. Kris Tompkins was far more diplomatic yet held her ground. In public and even at the dinner table, Doug and Kris were famous for impassioned arguments, dismissing the other's conclusions and shredding one another's logic. Their battles were notoriously intense, and yet when they were over it was as if they had changed channels and gone back to being civil. Friends viewed their ability to clash in public not as a sign of weakness but rather of the strengths welding their marriage.

Surrounded by open skies, lakes, and wild animals, the Tompkins home received a steady assortment of friends and visitors from abroad as Doug and Kris labored on their new national park project. They also spent days at a time hiking, climbing, or flying above their new home. When Doug spotted a mountain peak that he suspected had neither been climbed nor named, he jumped at the chance for a first ascent. Together with Yvon they struggled to reach the summit. It was not particularly technical, but on their first effort Yvon and Doug were too exhausted to reach the 6,211-foot peak. Yvon's boots cracked and broke. "We should call this Geezer Peak," he joked.

Descending to the lodge, Tompkins announced that in honor of his wife he was christening the peak "Mount Kristine." Tompkins was sure the mountain was anonymous and therefore could be named. Chilean officials agreed with him. Among the gifts Kris received from

Doug, it was Mount Kristine that topped the list. "He put me on the map!" she laughed.

At least 90 percent of our movement in Chile and Argentina was in the air, maybe more. Nearly everything we did was based around flights.

Flying changed the way we looked at the land and, more importantly, how we understood landscapes, especially the complex topography of southern Chile. It taught us how to look at coastlines, the islands and their relationship to the continent. We could never have understood the ecology of the Iberá Wetlands without the hundreds of hours flying over it. The national parks that we've donated were, in large part, understood through knowing every peak, every fold in the valleys, every pond, lake, and watershed. It was from the air that the big-picture landscapes came into view. Flying was also the direct means to articulate to those who were skeptical, politicians and community members, why an area was so important and see firsthand how it would or wouldn't influence communities.

Of all of my experiences in the south, beyond our love story, the biggest single impact on how I see the world, how I gauge beauty, how I found my true self is, without question, through the thousands of hours of flying with Doug. Of the two near-death experiences I have had, the ones where you are steeled and prepared to go; one was flying, the other sailing in deep winter along the Chilean coastline. These things sculpt who we are like nothing else can. We even went flying to buy groceries.

Doug loved me to eternity, but if our bed had been large enough, his little Husky would have been tucked in with us.

—KRIS TOMPKINS

Living on the sheep ranch was a logistical headache. It took years to sell off the entire flock. But as the sheep herd was reduced over time, a more profound beauty sprouted. Fields of golden grasses sprang up.

"The amount of grassland recovery, the pace of recovery of this ecosystem, was faster than anticipated. Which isn't to say that it won't take hundreds of years more. It will," declared Tom Butler, the editor of *Wild Earth* magazine and a longtime editor of Tompkins's books. "But it has bounced back quickly. You drive through Patagonia and you don't see grass. Then you drive into the Chacabuco Valley and you see this lush, waving grassland and big herds of guanacos. It just feels right. It feels magical, heavenly. It's pretty amazing what nature can do when allowed to recover."

Chapter 15

The River Killers

This is a country that has been sacrificed. Socially and environmentally speaking it's a country of sacrifice. Chile has murdered most of its rivers. The Loa doesn't flow into the ocean anymore. Whatever water it has is sewage and industrial waste tailings. The Aconcagua is considered a dead river. The Maipo is a dead river. The paradigm: Chile is for exploitation. So, who is this gringo coming to tell us we must create National Parks?

—JUAN PABLO ORREGO,
Chilean environmentalist

When a journalist called Kris and asked "How do you feel that you're just starting off with this new park and they're going to build dams on the Baker River?" Kris had no idea what the reporter was talking about. Then she read the newspapers.

A group of corporations announced they had selected the heart of Patagonia for a $3.2 billion complex of a dozen hydroelectric dams. The largest proposed dam measured 240 feet high and would require nine years to build. The Patagonian landscape was going to be sliced by power lines, electricity generation stations, and a noisy, dirty, nine-year construction cycle with some 6,000 temporary workers—which in South America meant shantytowns, prostitution, garbage galore, and the destruction of a peaceful rural life. The project was called HidroAysén and would be the largest energy project in the history of Chile.

The dam project was a partnership between Colbun, a company held by the Matte family, one of Chile's wealthiest, and the publicly traded Spanish multinational ENDESA. And they planned to build the dam only a few miles away from Chacabuco Valley. The proposed dams would interrupt the Baker River, into which the Chacabuco River emptied, meaning that the entire watershed was about to be disrupted.

Kris and Doug were stunned. They had been outfoxed by ENDESA in the battle to buy Huinay, the parcel of land in the center of their Pumalín Park project further north. Now ENDESA was at their doorstep again, this time with plans to build a series of dams to harvest the energy of a dozen rivers.

ENDESA and Colbun announced they had lined up financing for the construction of the massive dams. Describing the project as a twenty-year boon to the local economy, the two partners predicted the dams would generate $120 billion in revenue. Once the dams and power transmission lines were installed, they claimed, their electricity would travel first to a plant near the town of Cochrane, then would pass through about 5,000 transmission towers that would zigzag across the country—one huge tower every quarter mile for a thousand miles. The consortium estimated annual production of 3,000 megawatts, equivalent to one-fifth of Chile's electricity consumption. In Santiago, hydropower was touted as "clean energy for the masses" and the distribution system hailed as a dazzling engineering feat to construct "the world's longest electrical transmission lines."

To acquire the land needed for the new reservoir, the consortium bought out dozens of landowners. An artificial lake they proposed to create would flood the valley across from the entrance to Valley Chacabuco. According to ENDESA and Colbun, since the electricity came from water it was sustainable and therefore renewable "green energy."

Three dams were designed for the Baker River, and two to siphon energy from the Pascua River. Under Chile's water code and the 1980 Constitution, water rights had been privatized and river water transformed into a commodity to be bought, sold, and traded. The Pinochet government had granted ENDESA privileged concessions and favorable conditions to own the water and also to profit from its movement.

As ENDESA and Colbun engineers finalized plans to pour cement right across the road from the entrance to Tompkins's planned park, ENDESA had every reason to believe it was a done deal. Public opinion polls showed 57 percent of the Chilean public approved of the dam idea, while hydro energy was seen in a positive light by most citizens.

In trying to figure out how to respond, Tompkins turned to his international allies—including Robert Kennedy Jr., who years earlier had helped form the nonprofit called Waterkeeper Alliance. Kennedy was appalled at the idea of the dam. "Every river is a masterpiece," he said. "Most people in the world will never see the *Mona Lisa*, but everybody would be diminished if it were destroyed."

Political hacks and Chilean power brokers alike knew the dam project had been announced only *after* long negotiations and a clear promise of government support. The government would guide the project through any and all environmental impact complications—that was tacitly understood. Three billion dollars in construction contracts meant sufficient graft among Chile's elite to schmooze federal, regional, and local regulators. The enormous structures required thousands of truckloads of cement, hundreds of suppliers, 5,000 temporary rental units, and a supply chain to please all the political parties back in Santiago. The proposed Baker River dam would flood unpopulated valleys, turning parts of Patagonia into valuable lakefront property.

Colbun and ENDESA had been gifted the rights to Chile's water and controlled roughly two-thirds of the nation's electrical supply; the HidroAysén project would be a cash cow.

This was not ENDESA's first dam project in Chile. President Eduardo Frei had supported ENDESA's project to build dams across central

Chile's majestic Biobío River years earlier, ending the classic rafting routes that Doug had shared with his family and Esprit employees in the 1980s. The Biobío dam projects in the 1990s were heavily opposed by local environmental groups, but they lost. Chile had barely returned to democracy, and after seventeen years of harsh military rule Chileans activists weren't yet used to openly criticizing their own government. The ragtag group of environmentalists who opposed the project faced an uphill battle. They brought their defense of the Biobío River to Europe, to the United Nations, and even to the International Monetary Fund's headquarters in Washington, DC. They nearly halted the project. But ultimately, the Chilean government approved damming the Biobío River.

In Patagonia, ENDESA held a far stronger hand for damming the rivers. There was little to no indigenous presence in the Aysén region, and the local population was sparse and impoverished. For a few million dollars they could buy out rural families living in the zone. For a few million more they could sell the plan to Chileans back in Santiago, most of whom had never heard of the Baker River or, for that matter, Patagonia. After their victory damming the Biobío, ENDESA thought it knew the playbook of Chilean environmentalists, and discounted any chance that the activists, whom they considered a bunch of hippies, could do any more than put up a bit of resistance that might delay the inevitable approval process.

If HidroAysén was approved, Doug told collaborators, the floodgates of development would open. Patagonia would drown under a wave of destruction sold as *progress*. Damming the free-flowing rivers would serve a death blow to the long-term strategy that Doug and Kris had developed to allow Chacabuco Valley to return to wilderness while at the same time transforming the regional economy from one based on resource extraction industries into one based on conservation and ecotourism.

Dams are almost always a boondoggle. If you look at the economics on them, they're ways of privatizing the flow of the river, stealing it from the public—they almost always are constructed with giant public subsidies. And the people in these river valleys—their lives are permanently destroyed because the river's gone, and their homes are flooded and they're forced to move. It's okay to do that if there's a definitive public benefit, but in virtually every case where you look at dams, that benefit evaporates under close scrutiny. In a true free market, natural resources are properly valued, and it's the undervaluation of those resources that causes us to use them wastefully. So, it's a publicly owned resource that is privatized using public dollars, and they're almost all involved in dramatic economic chicanery that is going to benefit a few billionaires, making them richer by impoverishing everybody else. They raise the standard of living for themselves by diminishing the quality of life for everybody else, and they do that by escaping the discipline of the free market and forcing the public to pay their production costs. If you show me a polluter, I'll show you a subsidy. I'll show you a fat cat using political clout—Doug understood that what we were dealing with was not free market capitalism. It was corporate crony capitalism.

—ROBERT KENNEDY JR.

With ENDESA challenging Doug and Kris in their own adopted backyard, the fight became personal. It was time to rally the Chilean environmental community, in which Doug and Kris had developed relationships forged during years of shared battles. "Doug called us all down for a meeting at Chacabuco to talk it over," recalled Peter Hartmann, a Chilean environmental activist. "He said, 'This cannot happen. This project is terrible. We have to do something and I'll help you. Let's launch the biggest environmental campaign that Chile has ever seen.'"

Doug asked Juan Pablo Orrego, who had been a leader in the campaign against damming the Biobío, and who was director of his own

Chilean NGO, called EcoSistemas, to develop a strategic plan. Orrego and a team of Chilean activists prepared a 116-page report that concluded with a recommendation for a legal entity known as the Patagonia Defense Council. This was a coalition that included environmental groups, tourism operators, local citizens, and some sympathetic politicians. "Many of the NGOs objected. They said that allowing politicians to be part of it would tarnish the process," said Hartmann, who was on the council. "But we did it anyway. If we tried to protect every ego in the environmental movement that wanted to protect their purity, we were not going to achieve any objectives."

Tompkins was incensed at the idea of dams in Patagonia. He told the team that they needed to stall the approval of the dam. He was certain that the future of energy production was migrating to alternatives. New technologies, he insisted, were lowering the price of producing solar panels and wind turbines. The price for these nonconventional renewable energy sources, Tompkins said, was poised to drop. Alternative energy sources were just around the corner. "Doug had a wider vision of the world than we did," said Hartmann. "His global vision was a big asset."

With activists fanning across the country to support the largest environmental uprising Chile had ever seen, Doug took control of the messaging, specifically the look and feel of the media campaign opposing the dam. He knew that damming a river to produce electricity was controversial in any corner of Latin America. In Chile, it was a declaration of war. Many of Chile's young environmental activists had cut their teeth battling to save the Biobío. It was a bitter loss. They had lost a masterpiece, and there was a desire to even the score. A young activist working with EcoSistemas coined a three-word slogan for their campaign: "Patagonia Without Dams." In Spanish, the equivalent phrase, "*Patagonia sin Represas*," was romantic, and catchy. It crystallized the notion that Patagonia was a wild paradise, even if most Chileans couldn't find it on a map of their own country.

Organizing a nationwide opposition movement against the largest energy project in Chile's history was a challenge. Arguing against dams was in many ways seen as an affront to the nation's free-market development model, which was based on exporting raw materials including wood pulp, fish meal, and raw copper. Chile was not investing in value-added processes like turning the wood into furniture, the fish into fillets, or the copper into pipes. The post-Pinochet government, known as *La Concertacion*, had consistently proclaimed that the duty-free, export-everything-to-the-world strategy would continue. Protesting the development model was seen as both anti-investment and anti-Chilean.

Doug had an idea: What if he focused not on the dams but rather on the power lines? He considered the 1,200-mile-long electrical transmission lines a scar—as ugly as a clear-cut and many times longer. And that concept, that image—a 1,200-hundred-mile-long scar, gashing his beloved Patagonia and cutting through half the country—became the cornerstone of his media strategy. What if he made a ghastly collage? What if he and his team showed the power lines defacing every corner of Chile? Although the proposed electrical lines traveled nowhere near Torres del Paine, Patagonia's most famous national park, Tompkins hired a local advertising team to mock-up a photo with power lines slashing the view of its famous spires. Many of Doug's allies doubted him. "It's not even true!" they griped. "Don't worry," he replied with a grin. He was betting that an image so distasteful would stimulate a nationwide appetite for beauty.

The ad was placed at the Santiago international airport, and also on a billboard near the Chilean Congressional building in Valparaiso. Chile's power elite couldn't help but notice it. A fiery controversy exploded. How could Tompkins say that HidroAysén was going to scar the views of Torres del Paine?! Executives for the electricity consortium were furious, insisting that electrical lines in front of Torres del Paine was flat-out unthinkable. The ad wasn't fair, they howled. The pro-dam forces slid right into Tompkins's trap. The Patagonia Defense Council was ready with a reply—"If it's not okay there, it's not okay anywhere in Patagonia." "The thing is, that's a metaphor," said Juan Pablo Orrego, an architect of the anti-dam messaging. "And the guys from the company, they reacted, so the ads were very successful. Douglas was amazing. It made people think about what a power line does to the landscape."

Instead of talking about the dam, or talking about the possibility of lower monthly electric bills for millions, Tompkins defined the dam as a referendum on beauty. He then suggested that the advertisements include a mock-up of the power lines crossing in front of the statues of remote Easter Island with the slogan "Here It Would Be Inacceptable—in Aysén Also."

Delighted that his bet on beauty was paying dividends, Tompkins poured hundreds of thousands of dollars into a series of ads. He hammered home the message that HidroAysén was sullying the nation's international image. "We never focused on John Q. Public. This was oriented to decision makers, to senators, to public officials," said Hernan Mladinic, the sociologist working on the campaign. "That's why we spent money on big billboards leaving and entering the airport. We wanted to influence those kind of people."

ENDESA knew how to handle powerful Chilean politicians. They co-opted or hired them. Among all the companies in Chile, ENDESA was among the top ten that made political contributions to the center-left coalition that governed Chile following the dictatorship. To stay tight with the extreme right-wing, ENDESA hired Pinochet's lawyer, Pablo Rodriguez Grez, a key intellectual ally of the dictatorship, as general counsel. He immediately accused the anti-dam activists of being violent fanatics. "These people are ecoterrorists!" Grez declared. As a longtime professor of law at the prestigious Universidad de Chile law school, his opinion mattered. In reality, it was Grez who was the suspected terrorist. Before the Pinochet dictatorship, a paramilitary group that he founded—Patria y Libertad—allied with CIA covert operators to sabotage the Salvador Allende government. Their activities included a bit of "accidental murder" along the way. Patria y Libertad was the CIA's conduit to sabotage Chilean democracy, and Rodriguez Grez was one of the key operatives. Thirty-five years later, Rodriguez Grez was not in the streets stirring up trouble; instead, he was able to speed-dial a fascist, should he need their services.

State-sponsored assassinations and torture were no longer possible in post-Pinochet Chile, though the latent threat of violence had often worked to paralyze activists. "They ransacked our offices. We had an old house with a thick wood door, and then the owner had put an iron grate [on it] with a big padlock. It got yanked off," said Orrego, describing an incident during the unsuccessful battles to save the Biobío River. "They had to use a big four-wheel-drive truck or something to do that. There was nothing left inside: computers, telephone, fax, everything was gone. In the morning when we came in, they had sprayed graffiti with dollar-bill signs everywhere that read 'Stop Proselytizing, You Bastards!'"

Patricio Rodrigo, the head of Chile Ambiente NGO and a strategist on the pro-Patagonia campaigns with Tompkins, had shots fired at his office. "I got horrible phone messages that said, 'We've disappeared people for a lot less than a $3 billion dam project.'" As ENDESA and its allies fueled a backlash against the environmentalists, pickup trucks and mining vehicles in the region began to sport fresh bumper stickers that read "Patagonia Without Tompkins!"

As the heat on Tompkins grew in the Chacabuco Valley, on April 30, 2008, his world was further rattled by a 5.4 earthquake. The quake's epicenter was in the heart of his planned park in Pumalín. The rattling lasted six seconds and caused little concern among his staff or the residents in El Amarillo and nearby Chaiten. Chileans are accustomed to strong earthquakes, and it was in Southern Chile where scientists in 1960 recorded the strongest quake on record: 9.2 on the Richter scale.

Although it was common for the ground to shake, tremble, and rattle, residents in El Amarillo, Chaiten, and Puerto Montt—even across the border in Argentina—were baffled. The movements stopped for five or six hours and then, with a shuddering *Bam*, a tremor alerted animals and humans alike that an epic movement was afoot. A humming rattled from inside the earth; some residents described a cacophony of eerie shrieks. So many dogs barked through the night that residents barely slept. Aftershocks became so frequent they began to feel like a preview not a sequel. On May 1, eleven short tremors rippled the parklands in and around the Tompkins headquarters. Earthquake sirens wailed through the night as an atonal symphony groaned from underground. One survivor later described the subterranean soundtrack as "galloping horses mixed with screams."

By the early hours of May 2, earthquake tremors peaked at twenty per hour. Local residents became more fearful and described the shaking as so violent they feared the ground beneath their feet might split open and swallow the entire town. Virtually no one in the Chilean government or in remote Patagonia realized that in the year 7420 BC a nearby volcano had exploded, sending lava, ash, and destruction across the river delta—now the center of town for Chaiten and its 4,600 residents. Fewer still could be expected to understand that after 9,400 years of dormancy, a volcano in the heart of the Pumalín nature sanctuary was awakening.

At 3:40 a.m., on May 2, 2008, a massive explosion punched a hole through the side of 2,800-foot-high "Chaiten Hill" as it was marked on most maps. Throughout the pre-dawn hours and into the early morning, a plume of grey ash rippled 10,000, then 20,000 feet and finally 50,000 feet into the atmosphere. Burning cinders, blocks of rocks, and millions of tons of particles spouted from the funnel. By dawn, townfolk gathered by TV sets as the governor and newscasters described "The eruption of Volcano Michinmahuida."

Hearing the news while he and Kris were in Buenos Aires, Doug glanced at an image of the explosion and instantly realized the governor and broadcasters were completely wrong. Studying the landscape during thousands of hours flying at low altitude, Doug had developed a photographic memory of Patagonia using notable features like inlets, fjords, and volcanos for guidance. Michinmahuida was his volcano, on his property, and had a postcard-perfect-shaped volcanic cone. What he was seeing in the media was a warped hillside. Whatever is exploding, Tompkins told his staff, it's not Michinmahuida.

Back in Chaiten, the eruption was so nearby that residents thought they might soon be cooked alive by lava. Heat from the eruption melted snow, and sediment from the volcano poured into the river Rio Blanco bordering Chaiten. Within hours of the explosion, the Rio Blanco mutated into a landslide of volcanic ash, sodden dirt, roiling sand, uprooted trees, and noisily rolling boulders the size of a pick-up truck. Gushing down the mountainside, the river changed course and through the center of town. It stripped away Chaiten's main street, one of the few that was paved. Watching a webcam monitoring his Chaiten offices, Tompkins saw water trickle in, then pool, and when it was halfway up the tires on his pickup, he realized that a catastrophe was near.

Emergency evacuation plans were invented on the spot. Fishing boats were packed to the rail with residents. No suitcases. No pets. The docks were soon a cacophony of howling dogs, wailing children, and harried citizens. A gray river the consistency of cement washed away entire blocks of the village center. Swirling waves of sand and ash buried homes under tons of debris. In video footage taken the day the volcano exploded, Tompkins is shown taking a glance at one of the photos of the volcano plume, making a very quick, and accurate, estimation of the altitude it reached. "That's way up there, 20,000 meters," he said. "This looks like a nuclear bomb went off."

With his regional office flooded and his visitor center demolished, Tompkins lost the precious momentum to finally create Pumalín National Park—gained after seventeen years of struggle. He was stunned.

Was he fated to be the man who almost made Pumalín National Park? Four months earlier, he had inaugurated the Pumalín visitor center. Now much of his park infrastructure was buried, isolated, or covered in ashes.

During the ensuing days, clouds of thick ash washed across the zone, disrupting airplane flight paths as far away as Uruguay. The new park entrance had just been christened for the growing stream of backpackers, VIP nature safari operators, and potential donors. The eruption changed everything. Tens of thousands of acres of forest were affected by the explosion. Buildings were buried, their roofs collapsed from the weight of the ashes. Green pastures were painted gray, as if cement had been poured across the landscape.

Flying over the park, the volcano still smoking, Tompkins admitted the volcano "caused us all sorts of headaches and lots of money. It's changed the way we're doing things. We're having to postpone or cancel projects so we can pay for the damages."

In addition to his full-time offensive versus the dams in Chacabuco Valley, Tompkins doubled down on his commitment to El Amarillo, a tiny village that included many families with whom he'd worked over the years. Doug and Kris spent less time at their Reñihue ranch and, increasingly, slept near the new entrance to the proposed Pumalín Park in El Amarillo. To better understand local sentiments and glean insights into his reconstruction projects, Doug accepted the vice presidency of the El Amarillo Neighborhood Council.

The volcano upended Doug's carefully laid plans for Pumalín Park. Entire regions of the park that had been restored, cultivated, and practically manicured were now destroyed. Instead of polishing up his visitor's center, he had to start from scratch.

Doug was increasingly stressed and frustrated. He needed a change of scenery. Tompkins had been talking for years about joining Paul Watson of the Sea Shepherd, to join other activists in confronting Japanese whalers. But conservation work in Patagonia and Iberá took priority. Kris pushed. She told him to get on the boat, that he would regret not going. When he agreed to finally go, Kris was relieved. Watson called it Operation Musashi, after a famous Japanese swordsman-philosopher immortalized for using two swords simultaneously as he battled legions of enemies.

Chapter 16

Operation Musashi

You're at the end of the world; it's just you and the whalers. There were some big, big confrontations. Ramming ships. We threw stink bombs, we threw paint bombs, we shot water cannons into the exhaust pipes of the ships. It must have been exciting for Doug. He definitely had this mischievous look on his face.

—WIETSE VAN DER WERF, founder of Sea Rangers, a Dutch ocean conservation group

In December 2008, Doug Tompkins hefted a duffel bag aboard an old ship at port in Hobart, Australia, and reported for eight weeks' volunteer duty on a mission to confront Japanese whalers in the Antarctic Ocean. He was sixty-five years old, by far the elder statesman of a crew that included a Dutch violin-maker, a retired US Navy officer, and the bearded captain, Paul Watson. For a decade Tompkins had funded Watson's antiwhaling navy, yet this was their first mission together.

Operation Musashi's goals were to disrupt the Japanese whaling fleet and thus put an end to plans to kill minke whales and fin whales. Because whale populations were so decimated, the International Whaling Commission put exact limits on the slaughter. For 2009, the kill was set at 732 whales. The Japanese government authorized the hunt under the excuse of "scientific research," but this annual whale slaughter always managed to supply fresh whale steaks to restaurants in Hong

Kong and Tokyo where the traditional glass of sake was served with a side of whale blubber. For the activists, one whale was too many. As the 2008–2009 Japanese whaling season began, Tompkins volunteered his time and also shelled out a quarter million dollars to pay for the ship's diesel fuel during the months-long hunt.

Captain Watson zeroed in on the Japanese processing ship the *Nisshin Maru*, the nerve center of the multiship operation. If he tracked the ship, he could then send commando missions using inflatable Zodiacs to interfere with the harpoons, destroy the nets, and foul the ship's propeller. It was a dangerous mission with little chance of success. "We, with one single ship, have to find a fleet of six in an area of water that is over a million square miles," said Peter Hammerstedt, the first officer. "The best comparison is riding a bicycle and trying to find a caravan of RVs somewhere in America and there is no direct road to get there."

On his initial walk around the ship Tompkins noted the wobbly hand railings, an oil patch on the floor of the engine room, and a VHS radio powered by a hand crank. Designed for patrolling the rough waters off the coast of Scotland, the *Steve Irwin* was a 190-foot-long fishing boat without an ice-shielded hull. Bumping into a block of ice below water line could sink the ship in a matter of minutes. The closest rescue ship, a proper icebreaker like the *Polar Star*, was berthed in Sydney harbor, a five-day voyage away if weather was favorable in a swath of ocean that sailors dubbed the Ferocious '50, Satanic '60, and Savage '70s.

Few of the activists on the crew of the *Steve Irwin* had heard of Doug Tompkins, and he did little to reveal himself. A team from Discovery Channel was on board, filming the second season of Liz Bronstein's documentary reality show *Whale Wars*. As the film crew roamed the boat searching for characters, Tompkins ignored them. He avoided their efforts to interview him. He had joined the mission to meet the young activists and defend animals, and the last thing he wanted was more publicity. He was engulfed in enough of that back on land.

Paul Watson knew a thing or two about media relations. The media access he had given to Discovery Channel was strategic. The Sea Shepherd Conservation Society was based in Santa Monica, California—close enough to Hollywood to ensure that donations came in celebrity-sized chunks. Sean Connery, William Shatner, Christian Bale, and Pierce Brosnan had all donated cash to Sea Shepherds. The actress Darryl Hannah worked as crew. Captain Watson was committed to a long-term strategy for protecting marine biodiversity; he believed it was criminal for the Japanese whalers to use a scientific loophole in international whaling treaties to continue the slaughter of these magnificent creatures.

Navigating south from Hobart, Tompkins braced in his cot as the *Steve Irwin* rolled and pitched, banging through waves forty feet high. The first storm system they faced was the size of Australia and knocked out the gyroscope. The crew was queasy, several volunteers were vomiting, and then icebergs started showing up—first on radar, then through binoculars. A volunteer named Mal Holland was a ship captain in his own right, but on the *Steve Irwin* he worked as a deckhand. He described the dangers as "really mountainous seas, unqualified people, and ice."

Tompkins was assigned the job of quartermaster, meaning he worked on the bridge and used binoculars to scout the horizon for dangers and reported to the captain. His shift ran 4:00 a.m. to 8:00 a.m. As the ship journeyed farther south, the nights grew shorter until it was never dark. Tompkins scanned the ocean for lurking icebergs, which were few and far between when compared to the waves of wildlife flashing before his eyes. A sooty albatross with an eight-foot wingspan drafted the ship, floating for hours without flapping. Then the albatross would zigzag along a cresting wave, darting inches above the spray. Pods of humpbacks, fin whales, minke whales, and orcas surfaced within easy view. "There were so many amazing, beautiful whales," declared Molly Kendall, another deckhand. "They come right up to the ship. If we were a harpoon ship, it would be so easy."

While off duty in the rough Antarctic seas, Tompkins spent hours hunched over his laptop, pecking at the keyboard while thirty-foot swells rocked the boat. "Everyone has this general malaise about them in rough seas; even if you're not seasick, it saps your energy," said Mal Holland. "The last thing you want to do is look at a screen, or read emails, but that's what he was doing, and a *lot* of it." Through a weak satellite connection Tompkins shared architectural sketches with the team in Argentina, discussed ideas about remodeling the visitor center at Pumalín Park in Chile, and sent love notes to Kris—his beloved "Birdy." He was also finalizing the abandonment of his dearest organic farming projects. The 2008 financial crisis had sucked down his cash; money sinks like the 18,000-acre Laguna Blanca farm, a work of art from the air but a hugely expensive agricultural experiment on the ground, were no longer feasible. Tompkins was reluctantly cutting back his largest investments in organic farming. This allowed him to conserve cash for core projects. Tompkins lamented that he wasn't twenty-five years old and able to dedicate a full half century to his cherished agricultural projects.

Aboard the ship, his internet connection was frustratingly slow. It sometimes took an hour for a photo to download, and Tompkins paid dearly for it: the bill for his ship-to-shore communications totaled $25,000.

Approaching the coast of Antarctica after a brutal ten-day navigation, Captain Watson instructed the team to find the *Nisshin Maru*. Online sleuths attempted to hack the boat's email, track its GPS position, or at least unearth details from the crew's online posts. This season, the Sea Shepherds had brought a new weapon: a Bell helicopter. The gusts, the extreme cold, and the constant clouds and humidity made this among the most dangerous areas on Earth to pilot any kind of aircraft, but when the chopper lifted off the deck, within ten minutes it could run a survey that saved the *Steve Irwin* an entire day's search. The helicopter altered the rules of engagement. Visibility was best in the early morning hours, so the pilot often lifted off for a two-hour morning flight. On the third day of his search, flying an hour from the *Steve Irwin*, the pilot spotted a wake and then the shapes of ships. He had found the Japanese fleet.

The distance between the *Steve Irwin* and the *Nisshin Maru* was 100 miles. In the open ocean, at an average speed of fifteen miles per hour, it would take the ship roughly seven hours to bridge this gap. This ocean, however, was not navigable in straight lines, which doubled the time and made crossing that distance more arduous. Entire bays were frozen over. In the Antarctic summer, cracks opened up through ice sheets, creating tempting shortcuts. The same fissures could also squeeze shut as ocean currents swirled giant ice blocks. The Japanese were using the ice sheet as a shield against the Sea Shepherds ship. When spotted by the helicopter, the *Nisshin Maru* headed further into the ice pack.

On his way back to the *Steve Irwin*, the helicopter pilot found an open route into the ice, but squeezing through the passage would be tricky. Motoring at top speed, hot on the trail of the Japanese ships, the bridge crew missed a turn into open water. In order to have any chance of catching the whalers, Captain Watson decided to continue; they would navigate the ice field.

The ice became so dense around them that the radar operator could not distinguish ice from open water. The crew had to rely on volunteers with binoculars to avoid icebergs, many of which lurked beneath the surface of the water. "This is a very dangerous game that we're playing," said crew member Jeff Hansen to the TV crew. "We hit an iceberg we don't see, get a breach in this hull, and we're going down in seconds. There is every chance that we will not come back."

Within hours, Hansen's fears were realized—a screeching metallic wail vibrated throughout the ship. The *Steve Irwin* slid into an iceberg and took a punch to the aft section of the hull. The hull was not designed for ice-breaking and bent inward. As the winds topped forty knots, Captain Watson took the wheel from the second officer. Using a giant iceberg as shield, Watson steered the *Steve Irwin* to shelter in the lee. By morning the conditions worsened, ice sheets blew in, and they were trapped next to the iceberg. Watson steered the ship in circles, clearing a hole of ice-free water that despite his maneuvers continued to shrink. If the ice sealed, they would be surrounded by 100 miles of ice pack. Their only route out was behind them, but running the *Steve Irwin* in reverse ran the risk that the prop would shear off or become damaged by blocks of ice. Watson threw the ship into reverse anyway. But as the ice built up, he issued a drastic order: prepare the life rafts.

"We could sink right here," the second officer said, looking into the lens of the harried camera crew. Ordered to film down inside the hold, they refused. Instead they prepared to board a flotilla of Zodiac escape boats. Two volunteers lowered themselves into the bowels of the ship so they could inspect the hull. What they saw was terrifying. The ice was so hard it popped the steel hull back and forth, setting off a shriek. The paint cracked and walls and beams swelled as if they might explode. The volunteers scurried back on deck, while the engine crew was told "if the water starts to come in, stay here and stop it."

Watson ordered the helicopter aloft. He told the pilot to find an escape route—any direction, any possibility. As the tension mounted, Tompkins worked on the bridge. He scanned the horizon, reviewed the radar and the charts. As massive chunks of icebergs raked the hull, Captain Watson again took the wheel. "You've got to move slowly into the ice, keep the engine pressure to push the ice, and push through it," he said, speaking directly to the camera crew. "You just have to be patient."

Hovering above the cracked ice and slowly advancing, the helicopter pilot guided the *Steve Irwin* back to the open ocean. The chase was on and the crew prepared their arsenal: stink bombs, water cannons, and a pair of Zodiacs to buzz around and try to foul the props of the Japanese ships.

The Sea Shepherds launched the first Zodiac into the rough waters to get close enough to the Japanese, but by the time the boat was launched, the Japanese were miles away. As the Zodiac set out, fog reduced visibility and the boat zoomed away—in the wrong direction, losing radio contact with the *Steve Irwin*. "It's a volunteer crew, first time in the boat in two hundred meters of visibility," said one of the volunteers, Jane Taylor, a US Navy veteran. With the help of the helicopter the Zodiac was guided back to the mother ship and the mission postponed until the following day. Taylor worked with the Zodiac team to prepare them for the next time the Japanese were within range.

For a full week the conflicts escalated. As one Japanese whaler attempted to transfer a whale carcass to another ship, the *Steve Irwin* plowed into the middle. The two ships crashed in what the Japanese called "a deliberate ramming" and Watson dubbed an "unavoidable collision." He told the Associated Press that the situation was turning "very, very chaotic and very aggressive."

After each skirmish Watson led the Sea Shepherds in chasing the faster Japanese whaling fleet. The helicopter became more effective at helping the *Steve Irwin* find the Japanese whalers, and a route to them. The volunteers improved their launching of the Zodiacs, and on a later attack they managed to foul the propeller of one of the harpoon ships, but they were never able to catch up to the factory ship, the *Nisshin Maru*, to try to foul its prop and end the Japanese hunt for the season. The Japanese whalers were also ready with countermeasures. Every time the *Steve Irwin* managed to catch up, they hung netting over their entire decks, making it difficult for the Sea Shepherd's stink bomb attacks. They had their own water cannons ready.

Captain Watson kept chasing the Japanese until the *Steve Irwin* ran dangerously low on fuel. When he couldn't maintain the pursuit any longer, he ordered the ship to turn north, and begin the long and still-dangerous journey back to Australia.

By now, Doug had met all of the crew, and had even gone around with his camera and taken pictures of all the activists on board the *Steve Irwin*. Conversations in the mess soon spread, and several of the crew figured out that Tompkins was a conservation rock star in his own right.

Tompkins identified with the young activists aboard the *Steve Irwin* because they were committed activists. Determined to slow the destruction of nature and the slaughter of whales, they risked jail time and massive fines to take on the mission of sabotaging the Japanese whale hunt. It was adrenaline-infused work.

Tompkins continued to avoid the TV crew, which was searching for juicy soundbites, emotional drama, and scenes of tension among the crew on the bridge. He kept a low profile as he chatted with his watch mates. Sometimes there would be hours of silence, with Doug gazing at the whales, watching seals escape the orcas. Slowly the word spread around the ship that the old guy in the wool cap, the one up on the bridge, was "Doug Tompkins, the guy who started The North Face."

Mal Holland saw an opportunity and asked Doug to share his thoughts. Would he speak to the crew after his shift ended? Tompkins agreed. For over an hour he described the challenges of environmental activism, slowly letting on that he was not only the founder of The North Face and Esprit but also a leading figure in the field of wildlands philanthropy. "I had been on the ship for over a year. When Doug was on board, it was the first time that people sat down and had a discussion about what we were doing. He was very much a deep thinker. He had a methodical way of approaching business and projects," said Wietse Van Der Werf, a Dutch environmental activist. "We spent weeks and weeks at sea, so there was a lot to talk about. He had an artistic notion. He showed a lot of photos of the work he had done on farms, which was beautiful! So, besides being a thinker, and very much a doer, he was also an artist."

"The revolution starts with a first step," Tompkins explained to his growing green team. "How else would it work? We got to get started. If you look at all the big social revolutions in the world, it usually started with one person having the idea." Protecting nature, Tompkins passionately insisted, was an exercise in aesthetic revitalization. "Living in a beautiful world brings joy, it brings pleasure to have a healthy world. Personally, I don't see anything else that's as interesting to do."

Doug's lectures became a hit. The crew eagerly awaited the next meeting and he warmed to his audience, making jokes, telling stories as he often did at campfires with his own friends. "He would do an hour and a half lecture in the back of the bridge, and we would sit there and record it and listen," Taylor remarked. "He was teaching all of us, the younger ones, the greenhorns that wanted to know. What could we do more? What was it like in the past? What was the forum when you got together in San Francisco with all of your friends? What are you trying to do now?"

Tompkins was insistent that environmentalists would always be underdogs. "We have to be good, because we're up against a monster," he said. "We're the David-and-Goliath story. Our movement is small. But it has the right spirit, and we have a righteous cause on our side. But we're outnumbered, outmaneuvered, outmanned, and outfunded by the forces of development. The Japanese whalers are investing $70 million this season; we're investing $1 million. They've got the financial power, and we're a bunch of scrappy environmentalists down here. So, we have to be a lot smarter than they are. They have to spend 70:1 because they have to prop up their morally bankrupt position."

The 2008–2009 whale hunt was not stopped, but the *Steve Irwin* campaign slowed the slaughter. The Japanese didn't reach their quota of 985 whales, and were only able to kill 325. More than 600 whales were saved. The investment Doug made in Watson's Sea Shepherd direct action would be paying dividends to the Antarctic ecosystem beyond his life span.

When Tompkins disembarked from the *Steve Irwin* in Hobart, Australia, he carried with him the contact information for all the activist volunteers on board. He planned to keep in touch with these eco-warriors. A month after the trip, each received a poster in the mail from Doug, with a collage of photos of all the activists on Operation Musashi. Several of the crew asked Doug for donations for their own pet conservation projects, and Mal Holland journeyed all the way to Iberá to follow up with Doug, to interview him and pick his brain.

Wietse Van Der Werf saw Doug again in Amsterdam a few months after Operation Musashi. In that short time Wietse had founded his own direct-action organization, which he called the Sea Rangers Service. Using refurbished sailing ships and professional crews, his plan was to offer low-cost ocean monitoring services to the Dutch government, and even to some private enterprises.

"Doug had come to the Netherlands to visit the Queen. But something was wrong with his passport; it was going to expire in a few months, or something. They wouldn't let him into the country," remembered Van Der Werf. Tompkins told the Dutch immigration officials that he was there to meet the Queen, but they didn't believe him. It took one of the Queen's personal assistants calling to allow him into the Netherlands. After meeting with the royalty, Doug lunched with the young activist. "I had gotten an award for 50,000 euros, and I thought that if I could get another 30,000 euros we could have a budget to start this new organization," Van Der Werf recalled. "About halfway through lunch, I decided to put an 'ask' to him. 'Are you willing to make some kind of contribution?'"

Doug apologized, said he was very sorry, and told Van Der Werf he had really overshot his budget for the year, and couldn't invest much. But he *could* give the Sea Rangers Service 50,000 euros—would that help? "He made that contribution, and that along with the prize money kickstarted this whole Sea Rangers Service," recalled Van Der Werf. "It was him trusting our pragmatic approach, and that we were doing it for the right reasons. He was willing to take some risk, and he just said, 'Go do it!'"

Chapter 17

River Keepers

The further and further we go down the path of embracing the technosphere and perpetuating the dangerous worldview of human supremacy over the rest of life, the longer it will be until we return to what will eventually have to be an eco-local model of development. The model that we have now is a massive failure since it has produced the biggest environmental crisis in the last 65 million years (!) and ruined the climate. The facts are relentless and pitiless.

—DOUG TOMPKINS

Following the exhilaration of his front-row combat against Japanese whalers, Tompkins landed back in Chile, where he was immediately engulfed in the latest attempt by HidroAysén to sell the public on its increasingly controversial $3 billion hydroelectric dam complex.

HidroAysén insisted that Chile was running out of electricity. Soaring consumption meant blackouts were just around the corner. TV ads showed an entire football stadium going dark as part of a false-rumor campaign alleging that Chile faced an existential electricity crisis. To stab home the point, power supplies to the nation were being interrupted in ways that many Chileans considered not only suspicious but devious. To make the populace suffer just enough to align them behind the HidroAysén project, had the blackouts been deliberate?

Tompkins laughed at their logic, calling them industrial age dinosaurs.

He argued that GDP growth and electricity consumption no longer marched in step. Renewable energies were emerging as an alternative. A booming economy could also sport stable if not shrinking electrical consumption. "Hydro power?" Tompkins scoffed. "That's last century."

The electricity consumption projections ginned up by ENDESA were gutted by the 2008 financial crisis in the United States. The financial downturn, following wild speculations on the ability of US real estate to sell at ever-more-inflated prices, slammed the Chilean economy. Commodity prices for copper, cellulose, and fishmeal tanked. Chile's key exports were pummeled. HidroAysén executives nimbly exploited the economic downturn. What better way to combat sluggish GDP growth, they argued, than having a fresh infusion of several billion dollars? Thousands of construction jobs were dangled as the carrot to overcome the constant jabs from Chilean environmental activists.

When HidroAysén officials realized how successful Doug had become in organizing the media messaging for Patagonia Without Dams, they jumped at the opportunity to portray him as a radical environmentalist and even as anti-Chilean. They tried to make him the poster boy for the anti-dam campaign, ignoring the Chilean environmental activists. Tompkins rarely moderated his positions, making him an easy target for the pro-dam consortium.

ENDESA redefined its position not as pro-dam but as pro-Chile. Shouldn't the country have the sovereign right to develop? It painted Tompkins as a capricious foreigner who zipped around in airplanes while denying locals the right to watch TV or use washing machines. To finish off the job (and the river), HidroAysén turned to Burson-Marsteller, a consulting firm nicknamed the "Darth Vader of Advertising" for its willingness to defend polluters.

Burson-Marsteller's infamous client list included Three Mile Island, the leaky nuclear power plant, and the chemical company Union Carbide, whose factory in Bhopal leaked gas that poisoned an entire city and left 3,800 Indian civilians dead, plus thousands more with permanent lung damage. In South America, Burson-Marsteller ran image campaigns for Jorge Videla, the Argentine dictator accused of overseeing the murder of as many as 30,000 civilians in the mid-'70s. For $1.2 million a year, Burson-Marsteller prepared pro-regime copy, including a thirty-one-page advertising supplement in *BusinessWeek* promoting the economic opportunities being created by Argentine's iron-fisted ruler. The ad copy cheerily declared that "few governments in history have been as encouraging to private investment. . . . We are in a true social revolution and we seek partners. We are unburdening ourselves of statism, and believe firmly in the all-important role of the private sector."

The pro-dam forces created a new slogan: "HidroAysén: The Nation's Project." The response from Patagonia Without Dams was full-page ads mocking the "obsolete and destructive" idea that billions of dollars in corporate profits were necessarily good for the nation.

Tompkins launched his counterattack with a scathing review of HidroAysén's association with Burson-Marsteller. "If HidroAysén is so great, why don't they sell it on its own merits?" He lambasted the PR firm as the patron of lost causes. "Why do they have to hire agencies known for defending the indefensible?"

With presidential elections coming up, the Patagonia Without Dams coalition wanted to put HidroAysén on the agenda: every candidate would have to stake a position. HidroAysén's fortunes were boosted by the rising political power of a Chilean billionaire—Sebastian Piñera, number #765 on the *Forbes* 2009 list of the world's wealthiest individuals.

Tompkins saw Piñera as a wildcard. On one hand he was just the kind of wealthy businessman whom Tompkins specialized in publicly spearing and goring. Socially awkward and prone to colossal verbal blunders on live TV, Piñera also sported an independent streak far more modern than was typical of Chile's ossified political parties. He spoke excellent English, had studied in Boston, taught in the Chilean university system for years, and was an avid bookworm. He was also an adventure junkie. Piñera was a pilot and, like Tompkins, landed on beaches, backyards, or freeways when gas ran low.

After a day in Reñihue listening to Doug lecture on conservation opportunities, Piñera began his own search for wildlands to preserve. Doug and Kris tipped him off—the bottom third of Chiloé island was for sale. It was ideal for a park and, at 285,000 acres, large enough to ensure a conservation legacy. Piñera flew in, bought the land and announced the creation of "Tantauco Park"—a wildlands that would be open to the public and administered and owned by his own foundation. Tantauco was a mirror image of Pumalín Park, and Piñera even hired Carlos Cuevas, a key ally of Tompkins's, to shepherd the project forward.

Although not a prominent public figure to most Chileans, Piñera was infamous in the upper echelon of Chile's close-knit business circles as a dirty dealer. "His worst enemies are his former business partners," concluded an author who spent two years writing an unauthorized biography of the man worth $2.6 billion. Fellow executives described him as the kind of colleague who'd execute your plan with impeccable attention to detail—right after he stole it from you. Several asked pointedly, "Is Piñera a crook or does he just get very close to the line?"

As a young bank executive, Piñera had used contacts in the Pinochet dictatorship to avoid trial and a possible prison sentence for his role in a fraud known as the "Bank of Talca Scam." He went on the run from an arrest warrant for nearly a month, while his political contacts delayed his arrest and provided enough time to design an end run around justice. In the '90s, while Tompkins built his village at Reñihue, Piñera shrewdly gambled tens of millions of dollars placing bets on Chile's recently privatized telecom sector. He also purchased large stakes in the national airline LADECO, which was rebuilt into a profitable regional airline. Piñera was a visionary speculator and relentless worker, often clocking fourteen-hour days while surrounding himself with brainy and bilingual MBAs from Chile's Catholic University. Piñera was a passionate promoter of the free market, even arguing that "there is nothing the government can do that the free market can't do better."

In November 2009 Piñera trounced all candidates in Chile's presidential election and sailed to victory. With a billionaire, free-market president taking charge, the chances of stopping HidroAysén looked bleak. Then in the early hours of February 27, 2010, one week before Piñera took over the presidency, an 8.8 Richter scale earthquake rocked Chile, decimating the southern coastal towns near Concepción and leaving a path of destruction throughout central Chile. As aftershocks continued throughout the night, and as the outgoing Bachelet government botched the warning, tsunami waves killed 150 people on the coast, and life in Santiago was disrupted. Whatever plans Piñera had were shattered. His task was now rebuilding hospitals, schools, and thousands of homes destroyed by the massive earthquake. HidroAysén jumped into the chaos in search of a winning strategy.

The CEO of ENDESA's parent company, a powerful businessman named Pablo Yrarrázaval, donated $10 million in earthquake relief from the company coffers. Lest shareholders question the value of the disbursement, Yrarrázaval turned the donation into a PR show in favor of HidroAysén. First, he showed up at the presidential palace with a symbolic check the size of a coffee table, then pointedly asked the government to provide HidroAysén with "more objective" treatment and not to succumb to the "excessively large demands" of environmental protection legislation.

Along with showering the national government with cash and the rural Aysén community with student scholarships, swings, and seesaws, ENDESA went on a hiring spree. The company padded its payroll with former ministers and government employees to lobby its case. Tompkins, however, was crushing them in the court of public opinion, so ENDESA offered a succulent monthly salary (reportedly above $25,000 a month) to Daniel Fernández, a smooth, self-promoting executive running Chile's largest public TV channel, Television Nacional de Chile.

Fernández was the perfect political operator for the job. Long accustomed to negotiating the corporate and government cliques that financed and governed Chile's fledgling democracy, Fernández had what Chileans called "political wrists" that could twist in all directions. Under his stewardship, HidroAysén officials felt assured that they would regain public approval, which they assumed was all they needed to begin the nine years of construction and finally start operating the dams.

With Fernández at the helm, the HidroAysén campaign against Tompkins now included rumormongering stories to the Chilean press that Tompkins had sired illegitimate children in the rural south. Tompkins took the provocation personally and directed the Patagonia Without Dams advertising campaign to showcase the powerful men behind the consortium. He targeted the business leaders financing the destruction of free-flowing rivers for short-term profits. "Doug set forth the theme," recalled Elizabeth Cruzat, an accomplished designer working on the campaign. "We've got to unmask these people," Doug told his media team. "We have to talk about the real motivations behind the HidroAysén project. We've got to show their motives for building the dams."

Cruzat and her husband, Patricio Badinella, designed advertisements featuring the face of Eliodoro Matte, the most powerful executive of Colbun. They doctored a photo of Matte to make him look like a wolf. Then they created an image with the wolf's face wrapped in wool and added the body of a sheep. They gave Matte's son, Bernardo, similar treatment. The full-page ads were published in Chilean newspapers with the tagline "Wolves in Sheep's Clothing."

Chile's insular elite sat up and took notice, having never been called out by name before. "Doug had this correct idea that people like the Mattes are aware of things like prestige or reputation," said Patricio Badinella, the art director for the campaign. "And, rather than ENDESA, which was a huge anonymous corporation, Colbun was a company related to a well-known family name in Chile. Tompkins always thought we had to try to connect with them and show that what they were doing would erode their own reputation."

Tompkins always used a respectful tone and had met both Bernardo and Eliodoro Matte in person. He never forgot that he was a visitor in Chile and was careful not to make enemies gratuitously. He even invited the Mattes to visit Pumalín Park. When Doug called them up, he held spirited discussions as he outlined his defense of nature. To the Mattes he queried, "Do you want to be remembered in history as the family that destroys Patagonia?"

Instinctively protective of Chile's wealthy upper crust, Chile's leading newspapers refused to publish many of the Patagonia Without Dams advertisements. So Tompkins borrowed a tactic from Greenpeace. He held a press conference highlighting the banned ads and stirred up a media storm in which the focus of the entire debate was whether or not to publish headlines, including "When Foxes Guard the Chicken Coop" and "Patagonia—Not For Sale." Tompkins delighted in shining a bright light on Daniel Fernandez, the pricy PR consultant who was caricatured as a red devil with evil horns shaped like high-tension towers and an extension cord sliding out his butt.

The HidroAysén leadership team was stunned. Tompkins made them look ridiculous. National polling showed a surge in opposition to the dams. From the initial 57 percent in favor, the Patagonia sin Represas campaign masterminded by Tompkins swung sentiment sharply against the project. Now only one in three Chileans supported the dam.

Sebastian Piñera jumped into the fray, accusing the environmentalists of "being irresponsible and opposing everything." While talking to construction industry leaders he warned, "If we don't make the decisions now, we are condemning our country to blackouts next decade." The president's assertions unleashed yet another backlash. Academics, columnists, and historians alike were comparing Piñera to General Pinochet, who had famously reduced the future of Chilean government to the phrase "With me or with chaos."

HidroAysén officials next suggested that foreign meddlers were attacking Chile's natural resources. Banking on xenophobia, they suggested that the environmentalists were fakes and that really the anti-dam campaign was a front by foreign electricity companies seeking to steal a great business opportunity from ENDESA. "Daniel Fernández is desperate. They are paying him to carry out the project and he is not achieving it," mocked Sara Larrain, a member of the Patagonia Defense Council. "This is the desperation of someone who let himself become a mercenary for the company."

Tompkins continued to rattle the status quo. Every time the Chilean media—and increasingly the international press—reached out to him for comment, he lambasted the dam owners as nearsighted businessmen with no sense of national patrimony. "The passport is meaningless," Tompkins said, addressing criticism that he was a foreigner meddling in Chilean affairs. "It is really your behavior that determines whether you are a patriot. If you're trashing your own country, ruining the soils, contaminating the waters and the air, cutting down the trees, overfishing the lakes, rivers, and oceans, you're not much of a patriot. I see a lot of nationalists pumping their chests about being such patriots, and meanwhile they're trashing their own country!"

Tompkins even took the anti-dam campaign overseas. Newspaper and billboard ads ran on the side of London's double-decker buses, in which Tompkins mocked the Chilean government's plans to plunder Patagonia. "We were talking directly to the political class and telling them: 'We accuse you!'" said Cruzat. "We placed ads in England to embarrass them, to say to the world, 'Look! This is what they're doing.' For Chileans to be exposed like that was very shameful."

A myriad of different environmental groups now saw an opportunity to save the rivers and doubled down on organizing local opposition to the dam. Caravans of cowboys waving flags adorned with the Patagonia Without Dams logo garnered massive television news coverage as they rode horses in mass protests. A who's-who of Chilean environmentalists plus hundreds of enthusiastic volunteers joined the uprising. In many ways, Tompkins was reviving his marketing skills from the Esprit years. At Esprit he had infused the catalog shoot with adrenaline and a sense of buzzing excitement. Even customers felt like they belonged to a larger movement. Now in Patagonia, and throughout the nation, those same talents lit a fire in the hearts of thousands of Chileans. But instead of promoting disposable clothing and short-term fashion trends, Tompkins was now promoting nature.

When he first launched The North Face, Tompkins had stressed the beauty of the mountains, the lure of the outdoors. Now in Patagonia he was launching a campaign that once again appealed to the heart. "The prioritization of beauty is, in some ways, totally aesthetic," said Nadine Lehner, an executive assistant to Doug and Kris for six years. "And in some ways it's also this very nimble move to recognize what people gravitate toward and how to create a brand or create a feeling that people crave."

Tompkins knew precisely what motivated young people. Like the volunteers who donated their free time to pulling up weeds and fence posts at Patagonia Park and the hundreds who lined up to pose in the Esprit catalog, Tompkins, through his behind-the-scenes financing and strategy, now masterminded a unified campaign to save rivers and to protect Patagonia.

Along the 1,700-mile-long route of the proposed line of electrical towers, volunteers went property by property, door to door, informing locals about the reality of accepting a massive tower in their yard. "A lot of groundwork went into this," said Cruzat. "A lot of people talking to locals, saying, 'They're going to destroy the value of your lands if they build a tower here. You won't be able to sell it because you've got a high-tension tower in front of the house. And you won't be able to move because wherever you go, a tower will be there.'"

In a desperate move, Chilean legislators offered a customized law for HidroAysén exempting the company from environmental impact assessments. Under the proposed legislation, dams could be "fast tracked" in the name of energy security. Tompkins and his team went ballistic. They fired off ad copy, editing and designing into the wee hours of the night, and launched yet another advertising blitz, this one depicting the businessmen as fat pink pigs with evil blue eyes and a mouthful of US $100 bills.

The campaigns created a stir in Chilean society. Eliodoro Matte began long discussions with his son and heir, Bernardo. Was the investment worth risking the family's reputation and their most valuable asset: the family name? Was Tompkins perhaps correct? Was the Matte dynasty investing on the wrong side of history?

As the debate blossomed, demonstrations erupted in towns across Patagonia. Cowboys blocked traffic. Throughout southern Chile, a growing coalition supported the vision of Patagonia Without Dams. Even the salmon lobby, which abhorred Tompkins and was among his fiercest critics, fretted that the dams might damage lakes and rivers. They joined the anti-dam coalition. Ad by ad, march by march, industry by industry, the pro-Patagonia campaign surged. "With time, Doug started figuring it all out, and surrounded himself with people who knew how to manage politics in Chile," said the activist Peter Hartmann. "Doug realized that sometimes you just have to be patient, and the only thing to do is wait."

When the HidroAysén consortium sent a fleet of pickup trucks to roam rural Patagonia in a ranch-to-ranch public relations blitz, it didn't take long for the Tompkins team to respond. HidroAysén employees in the pickup trucks promised the rural landowners that the dams were a sign of progress, and that they would bring cash. The new lake, they swore, would not be an eyesore but rather a tourist magnet. Not only would the transmission towers be good for the economy but they meant fast cash to whoever signed up first. They tried to explain that the rural landowners were fortunate that instead of merely having the wires pass by, they would be handsomely paid just to let a simple tower be built.

Searching for a way to counteract the company's offensive, art director Badinella called up the offices of HidroAysén. He mimicked the drawl of a rural Chilean cowboy and asked the secretary at the company, "Whatdidyasay them company trucks looked like? I wanna be ready when they pass by." The receptionist was helpful. The trucks, she explained, were white, with a blue logo on the door.

Using that description, Badinella and Tompkins launched a radio campaign portraying the employees in the white pickups as hucksters on a mission to destroy the Patagonian way of life. The radio spots urged local farmers and ranchers to fight back. To defend their land. "Don't even open your door!" the ads suggested. "Don't let them on your property!" The seeds of a rural backlash were planted.

The Patagonia Without Dams campaign also hired songwriters to write rhyming ballads known as *payas* to ridicule the dam project. The *paya* is a Chilean country music style built on spontaneous biting rhymes, like slap-down rap. In *paya* competitions two *payadores* banter back and forth, insulting one another in a verbal duel that leaves the audience on the floor in laughter. "There are some formats of the *paya* in which you can insult people, while in others you must be a gentleman," explained Badinella.

The Patagonia Without Dams campaign even designed comic-book-style ads featuring stories in which Don Epifanio, a whitehaired and wizened Patagonian cowboy, discussed the false promises of the dams in a conversation with his horse. The pro-Patagonia campaign added two new phrases: "Destruction Is Not a Solution" and "Chile Says No to HidroAysén."

Still, HidroAysén moved forward. By an 11 to 0 vote, the environmental committee for the Aysén region approved the project. The green light from local regulators soothed the worries of HidroAysén executives—finally, they were harvesting the fruits of funding local economic development schemes, student scholarships, and promises of subsidized energy tariffs for the region. Their 10,500-page environmental impact study was proof that they had studied every angle and answered everyone. Nine days later, the consortium's victory lap was rudely interrupted by the largest public demonstration in Chile in over a decade.

A crowd of some 70,000 people marched through central Santiago. The demonstration stretched for a mile—from Plaza Italia to La Moneda, the presidential palace. The protestors marched peacefully to Piñera's office and delivered a letter. The Patagonia Without Dams campaign then lit the match of social activism. Long-simmering complaints and grievances were given a stage, and the anti-dam march helped unleash the demands of the Chilean people, which had been in a nearly thirty-year hibernation. Other social movements soon erupted with a hunger for justice and a cry to be heard. Piñera was weakened further with the eruption of another democratic uprising, this one led by high school students.

When Piñera declared that public education was "a market commodity" and should be priced accordingly, the reaction was instantaneous. Hundreds of thousands of high school students, led by college students Camila Vallejo and Giorgio Jackson, shut down the nation's schools for nearly a year with sit-ins and marches. Tens of thousands of students occupied their high schools, living inside and raising money through selling tickets to live music concerts presented pro-bono by sympathetic bands. Piñera and the elite felt dumbfounded. What was happening to the social order?! How had teenagers become revolutionaries?

Unlike their parents' generation, these Chileans born after 1980 had not faced the bloody torture squads of General Pinochet and his feared DINA and CNI secret police. Not knowing the precariousness of Chile decades earlier, they felt little loyalty to the development model that had undoubtedly lifted millions from poverty. The teenage protesters provided even more bodies for the anti-dam campaign.

The rowdy yet peaceful street protest demanding "Patagonia Without Dams" also reinvigorated a historical Chilean belief that tens of thousands of citizens taking to the streets could create a more equitable democracy. Throughout the nation's history, rebellion has been a well-honed tool in the fight for social justice. By 2010, few marchers feared anything worse than tear gas, a light beating, or brief imprisonment. Tortures and disappearances were a thing of the past. As Chileans became emboldened, sparks of protest flared.

Organizations that were not part of the Patagonia Defense Council launched their own anti-dam demonstrations. "People were coming to us and saying, 'What time is the demonstration in Puerto Montt?' And I'm saying, 'Well, I don't know. I'm not organizing it,'" said Mladinic. "That was the moment we realized this was a big national wave, and unstoppable."

The government's support for HidroAysén wilted. The project was suddenly bogged down by the invisible workings of bureaucrats who understood the project was, politically speaking, dead in the water. Political channels that HidroAysén officials had greased in preparation for the approval of their multibillion-dollar extravaganza were suddenly clogged as a political tide washed back upon them.

"Doug worked like crazy on the campaign, but he had fun," said Badinella, the art director. "I've worked for a lot of companies with very sophisticated clients that knew a lot about communications. But to have the combination of abilities that Doug had? To be able to know what you're talking about? Know about the media at your disposition to achieve your goal? And then to trust in the people you're working with and leave a door open for them so that they can create ideas? That's difficult to achieve. If you ask me if, in seven years working with Doug, if he ever messed up, I'd say probably not. And as everyone who works in advertising knows, generally when the client sticks their nose in too much, they usually screw it up."

While the HidroAysén consortium continued behind the scenes to try to move its project forward for a few more years, the political support never returned, and it got stuck in an endless environmental approval process. By 2014, the project was stalled, and then HidroAysén's backers quietly abandoned the project a few years after that. "The Patagonia Sin Represas campaign became something cultural," Cruzat insisted years later. "Before that, if you asked someone in Chile 'Where are you from?' they might answer 'from the south.' Today people say, 'I'm from Patagonia.' There are Patagonian restaurants, Patagonian food, arts and crafts from Patagonia found throughout Chile. All that has occurred in less than ten years."

Down in Patagonia, Kris and Doug Tompkins were delighted. The pro-Patagonia campaign cost $6 million. It was a small fortune for any environmentalist and could have funded the building of many miles of hiking trails, hundreds of acres of organic farm upkeep, or thousands of Alerce seedlings. But Doug found the money exceptionally well-leveraged: with $6 million the Patagonia Defense Council had stopped a project budgeted at $3.2 billion. By the end of the seven-year battle, Doug had drowned the dam in bad publicity.

Part IV

Chapter 18

Puppet Shows for Parrots

Unless we learn to share the Earth with all the other creatures on the planet, our own days are numbered. And that means demanding of our governments to make biodiversity conservation a priority. The primary means to this end will be more protected areas and, best of all, more national parks.

—DOUG TOMPKINS

Piloting just 100 feet above Iberá as he surveyed his vast acres of property in the remote Argentine wetlands, Tompkins marveled as hordes of caimans basked in the humid semitropical sun and herds of capybaras (the world's largest rodent at 100 pounds) trotted through clearings on the floating islands. The abundant birdlife made piloting hazardous. A stork ripping through the single engine of his plane could send him crashing into the wetlands.

As he piloted at low altitude or kayaked through reeds, Tompkins eyed the wildlife with the eyes of a naturalist and the heart of a rebel. He wanted to buy it all and then let nature recover. The dozens of lawsuits in defense of the Iberá wetlands that he had spearheaded were succeeding. Forty-seven court rulings were issued—each in favor of environmental protection. Not only were the waters of Iberá flowing again, but wildlife populations were increasing. Restoration had begun.

Tompkins savored the legal triumphs and the early signs that habitats could bounce back far more quickly than anyone predicted. Yet he knew that it was a fragile victory. Kris, Sofia, and Marisi brainstormed: How could they encircle this wonderland with a moat of defenders? Much of the solid ground was owned by a few elite and wealthy owners. The vast majority of the local population, poor and living outside the formal economy, was clustered in a dozen small towns that suffered a constant epidemic of "brain drain." The young and the educated were abandoning the region for opportunities in larger cities including Buenos Aires and Rosario. Local culture was fraying.

After ten years battling with powerful economic interests in Chile, Doug and Kris had learned the importance of team building. And the more they talked with locals in Iberá, the more they realized these Correntinos were political orphans. Like the Chilean settlers in the rain forest on the other side of the Andes, these rural Argentines felt extraneous to the decision-making process. Ignoring their needs, desires, or demands carried virtually no political cost. Rarely were they courted or counted. "Some of these small towns have almost no budget, or even much participation. The province is very poor, and these were some of the poorest municipalities; just that we contacted them and met with them made an impression," said Heinonen, the Argentine wildlife biologist. "We invited all the mayors and their wives to Pumalín." Doug cooked a *curanto*, a traditional Chilean dish. He gave a tour, took care of them, and even flew them over the park. "Doug told them, 'This is what we are thinking.' The mayors saw all that and couldn't believe it. They started talking about a future," she reported.

Corrientes senator Sergio Flinta, who had clashed often with the hard-headed gringo, went along. Flinta had doubts about the trip to Pumalín Park. He feared the political fallout if it became public knowledge that he was fraternizing with the "enemy." Tompkins sweetened the offer. If Flinta would fly to Pumalín and spend a week investigating the possibilities, Doug would purchase and rehabilitate a waterfront parcel in Colonia Carlos Pellegrini. He would donate the land back to the town as a community campground. The deal insulated Flinta from charges that he had nothing to gain by meeting the *gringo loco*. For Doug and Kris, the campground was valuable as a cornerstone to their relationship with locals and the land. Upgrading a campground would serve to blunt arguments that the couple were just the latest greedy patrons coming to loot their riches and horde them for themselves. The campground would include boat docks, which provided public access to the wetlands, something local communities had long been missing.

In Reñihue, Tompkins strapped the burly senator Flinta into the backseat of his Husky and revved the engine. They hurtled down the craggy airstrip in his front yard, then Tompkins banked steeply up as he whirled into Doug's airspace—a world he knew so well that no one else could provide such a panoramic, informative, and terrifying tour. They circled above the pristine organic gardens, buzzed by the remodeled ferry terminal, nearly brushed the face of a waterfall, and photographed a sea lion rookery teeming with dozens of the playful creatures. At night, by a campfire, Tompkins and Flinta drank wine, argued, and laughed. Late in the evening a grave misunderstanding erupted into an argument and then a bitter shouting match. Tompkins had been pressuring Flinta to endorse not a *regional* park but a *national* park. Flinta stated that local passions burned with a desire to be independent of the authorities back in the nation's capital. There would be no national park. The men started screaming at one another. Their entire fragile alliance fractured over a single phrase lost in translation. The next morning the two men hardly spoke. Flying to the heart of Patagonia for a visit to the Chacabuco Valley conservation project, Flinta felt miserable. "The row I had with Doug deprived me of enjoying the most amazing landscape I had ever seen, which was flying from Pumalín to Valle Chacabuco, soaring over the Patagonian Andes of Chile from the air. There's nothing like it in the world. But I was ready for the trip to be over."

Landing in Valley Chacabuco, Flinta was surprised that Kris awaited him with cheer and food. Though she had been warned, she acted as if she had no idea about the blowup. They toured the organic garden and ate roasted lamb under copper-plated lamps worthy of a Buenos Aires opera house. "Kris silently took care of it. She even got me to cook in her house, all to raise my mood," said Flinta. "She raised our spirits with her warmth, her candor, and her love. That's why I always say that you can't talk about Doug without talking about Kris."

Flinta overcame the bitter campfire argument and returned to Argentina mesmerized by the sights of national parks that he felt seamlessly integrated local tradecrafts and traditions while keeping out the chain hotels, the cheap furnishings, and the consequence of short-term thinking. No matter what, he still had that waterfront campground for his constituents.

Protecting the wetlands of Iberá was just the first salvo. Rewilding the park with native animals could take decades. Neither Kris nor Doug assumed they would be able to finish the rewilding efforts in their lifetime. Yet they had a clear dream and a rough sketch of how to begin.

Doug wanted to start with the jaguar. It was majestic, magic, and as an apex predator could be hugely important to the overall stability of the lands. Kris fought the idea. It was too soon. The Argentine team pushed for an alternative: Why not start with the giant anteater? Through the foundation that they managed, Kris and Doug agreed and budgeted tens of thousands of dollars to first build and then staff a rehabilitation center for the giant anteater, perhaps the most bizarre-looking animal on the continent.

With a snout as long as a vacuum cleaner tube and a retractable tongue that carries barbs and sticky glue to catch ants, the anteaters filled an important niche in the ecosystem. The animals have a three-foot-long bushy tail and claws that look like they belong on a dinosaur. An anteater appears gentle yet is capable of gutting a person. With a keen nose but limited eyesight, the giant anteaters were killed, leaving behind clutches of orphaned anteater pups. "The grand majority of the anteaters that come to Iberá are orphans that have been rescued because their mothers have been killed or hit by cars in traffic accidents, during hunts or in fights with dogs," explained Alicia Delgado, a lead biologist in the re-wilding initiative. "When people find the babies they take them home."

Local residents found that baby anteaters made cute house pets. They were adorable babies that needed to cuddle and be nursed for eight months. But when fully grown, the anteaters were untenable as pets and were seen as a nuisance and a dangerous house guest. "We could not have reintroduced a more appropriate species. It's weird looking. It's shocking. Harmless, for the most part, unless you corner one," explained Kris. "The anteater is so adorable! Orphans are coming in one door to quarantine, and out the back door as we reintroduce them. They have names, and people follow them. They're in puppet shows, they're on TV, they're famous. The anteaters are rock stars!"

Following the successful reintroduction of giant anteaters, veterinarian Gustavo Solis, a leader of the Tompkins couple's rewilding team, proceeded to investigate the feasibility of reintroducing other species, including the pampas deer and the initial moves to reintroduce wild jaguars. Solis committed his time and energy to understanding the intricacies of bringing the wild jaguar back to the swamps, wetlands, grasslands, and marshes of Iberá. Solis knew that this vast landscape was ideal jaguar habitat—and for hundreds of years, the jaguar had been revered by the native Guarani people. The word jaguar is derived from the Guarani term "yaguarete," which is how most Argentines call this species. Now the revered cats were nearly gone. "In Argentina there are only three separate jaguar populations, and we estimate that there are 200 jaguars. Every territory that is lost is a threat to the species," said biologist Maite Ríos Noya. "The disappearance of the jaguar would be a catastrophe. I don't even want to think about jaguars as we think about dinosaurs—as an extinct species."

When a student came to Ignacio Jimenez and asked to study the jaguars, he politely told her that it was impossible. There simply weren't enough jaguars left to research. Jimenez suggested that she study popular attitudes about jaguars. If there was going to be any chance of reintroducing the symbolic feline to the Iberá grasslands, Jimenez knew that it would take broad community support. And that required outreach. Jimenez needed the raw data, so he sent the student off, and when they analyzed and quantified the results of 432 interviews, the entire team was stunned.

Locals loved jaguars. They identified with the cat as a regional symbol of their warrior culture and rebel nature. For Doug the synchronicity was a welcome tailwind. Instead of battling against local misunderstandings as had been the case so consistently in Chile, here he might be able to tap into a latent love for the very species he so passionately wanted to rewild. Doug authorized construction of massive pens to hold the jaguars—and as local construction crews commenced they were hammered by unusually strong rains, turning the project into an epic, mud-soaked challenge that lasted two years. Meanwhile biologists, veterinarians, and members of the Tompkins team visited India, Brazil, South Africa, and Spain in an effort to understand the complexities of big cat breeding.

Despite misgivings from Doug and Kris, the rewilding team also pushed to bring back the red-and-green macaw. Hunting, poaching, and smuggling had eliminated the local macaw population a hundred years earlier. In wild corners of South America, further from human encroachment, the parrots still survived in native habitats, but locally the parrots were only found as pets. When word filtered out that the new gringo owners would provide homes for parrots, donations trickled in. "They were macaws but they acted more like people than birds," laughed one veterinarian. "We had to teach them to fly."

Reluctantly Doug and Kris gave the go ahead and hired staff to feed, care for, and ultimately prepare the rehabilitated macaws to live in the wild. The project failed spectacularly. The parrots learned to fly in such a limited way that when released they were promptly snarfed up by local predators. On the first release, all died except one. "The macaws were attacked by other birds after they were released," said Nicolas Carro, a veterinarian on the team. "Birds are very territorial, and birds like the Strange-Tailed Tyrant and other smaller birds treated them as intruders. The macaws ended up living on low ground in the spiny plants and were killed by foxes and alligators. They were like office workers stuck into a professional football match—quickly exhausted."

Members of the rewilding team asked around. Hollywood animal trainers prepared animals for movies, so maybe they could help train native animals to live in the wild? Through his network, Tompkins heard of an expert bird trainer, an Argentine said to be the best in the film business. But Fabian Gabelli's resume lacked one key skill set: he'd never trained birds to be wild. In fact, he had spent years domesticating parrots and getting them to perform on camera. Faced with the request from the Conservation Land Trust, he was asked to invert his skills. Could he perhaps "undomesticate" a macaw?

Gabelli leapt at the challenge. Right away he understood the failure of the Tompkins team's initial efforts to reintroduce the macaws. The Iberá wetlands are a vast ecosystem with small islands of forests and trees that allow the parrots to forage and nest. With weak chest muscles—a result of living in cages without the opportunity to fly—the birds tired easily, so when they landed on the marshlands they were easy prey. At first the Conservation Land Trust team didn't understand Gabelli, and there were a number of arguments. Sofia said, "Fabian, birds love to fly!" And Fabian explained, "No! Birds fly because they don't have any other way to solve their problems! It's a very human viewpoint to think that birds love to fly just because we would love to be able to do that. In nature, birds fly to find food—that's their life. Flying is just the means to do that."

Gabelli put the macaws on a strict food and exercise regimen. The Tompkins team built a training center for parrots consisting of an eighty-foot-long enclosed track with bird feeders at each end. When the birds completed the circuit, they were rewarded with food. Reviewing the failure of the first attempt at rewilding, Gabelli noticed that the birds were fed communally, all from the same feeder. The dominant birds ate all the fruit, and subordinate birds had to make do with the leftovers.

Neither group received a balanced diet. So the vets and biologists separated the birds into groups and divided meals into individual portions. Within three weeks all the birds were developing their pectoral muscles, which they'd need for extended flights.

Marianella Masat, a biologist working on the project, was bedeviled by setbacks as her team designed artificial macaw nests. The initial nests, made of plastic, were too heavy, and when she switched to wood-based nests, these were colonized first by bees, then by owls.

The veterinarians knew that macaws eat a variety of native fruits and seeds, and that while flitting, flying, and defecating on the move, they disperse those same seeds over hundreds of miles, helping to regenerate ecosystems. Biologist Jimenez called the birds "forest builders."

As the birds grew stronger and could fly faster, Noelia Volpe, an Argentine biologist working with the birds, added more obstacles to the course, and more problems for the macaws to solve. "We train their muscles so they have resistance when they fly, because before they had only flown in zoos," she said. To reinforce the bird's grip, poles of varied diameters were used to fine-tune their muscles. Individually tailored diets of fruits, seeds, and plants were introduced. Feeding stations set up along their training routes were set further apart and, with obstacles along the way, grew more difficult to locate. The times the birds took to complete the course, and the calories they consumed, were all recorded to measure their progress.

"The birds were treated like elite athletes," explained Gabelli. "They had to receive their calories for each day and complete exercises, with all the information about their performances recorded and studied. It was magnificent to be training animals for wild environments instead of something related to humans."

Extending the curriculum, Gabelli began teaching the parrots how to recognize predators. He built an amphitheater for his birds, in which the main stage was a puppet show depicting the dangers of life in the wild. Gabelli opened the show with a puppet of an eagle. The parrots didn't react. Having never been in the wild, they had no fear of—perhaps even no recognition of—the eagle as a prime predator. Gabelli tried a new tactic: What if the eagle attacked a parrot and at the same time he broadcast the warning cry of the macaw? "We staged an attack and the birds reacted like crazy, and moved away like they were watching predation in real life," said Gabelli, who forced the birds to watch the attack scenario over and over. Sofia visited while Gabelli was rehearsing. When he finished she said, "This is theater. I can't believe we're doing puppet shows for parrots!"

After months of exercise work, and puppet shows portraying the predators the birds would face, Gabelli had a group of macaws ready for release into the wild. Some of the birds would wear radio-tracking devices. "Before we released the macaws, we had this big campaign and provided a phone number. 'If you see one, call us and let us know.' It's the macaw hotline," said Kris. "And we got calls. People took pictures. Told us the macaw was 'about this big. He's in our backyard.' 'And this is the kind of tree he sat in!' People went nuts."

"You have to convince society that keeping natural ecosystems with integrity—which means keeping big animals that were lost, like bison, or wolves, or tigers, or cougars, or jaguars—makes more sense than other options," said Jimenez, who worked for years on the rewilding efforts in Iberá. "Once you know that, you use every tale, every story, every myth that you can induce from local and national society. If you use arguments and reasons that only resound with conservationist groups, you're going to lose. What do people really care about? They care about jobs, they care about pride, they care about hope, they care about culture, they care about patriotism."

After environmental nonprofits in Brazil and Paraguay offered male jaguars to commence the breeding program, Doug and Kris authorized construction of a massive Jaguar Reintroduction Center. The construction and plans of the JRC were on the scale of a metropolitan zoo.

Hundreds of acres were mapped and elaborate pens designed to allow jaguars to move from small veterinary pods to small breeding pens to initial release pens and then gargantuan 70-acre pens. There were so many structures that when flying into San Alonso the jaguar installations stretched for what seemed like miles across the flatlands.

Biologist Sebastian di Martino knew that the reintroduction of the jaguar was crucial to overall health of the wetlands. "Just as the return of gray wolves to Yellowstone National Park recalibrated whole ecosystems that had fallen out of balance, jaguars can restore these wetlands," he said. "Rewilding is also revitalizing the economy of small communities throughout Corrientes Province through wildlife-watching and related services. We have examples of small towns whose economy is based mostly on ecotourism and wildlife."

Di Martino knew that he needed to first release females since they have "smaller territories and disperse less." The strategy was to build a healthy, genetically diverse population. But given all the unknowns, he'd do what he could to keep the cats inside the borders of the park and as far from humans as possible. One incident. One setback. Or the death of either a jaguar or a person would set the project back years.

As they spent more time in the community, Doug and Kris realized the Correntinos tradition included a profound veneration for local wildlife. The annual carnival parade and celebration featured feathers and tropical colors. "A gaucho is kinda like a peacock; he likes to show off," said Sofia, describing how, together with Kris, she burrowed into local culture looking for a route to reintroduce and rewild animals native to the area while building upon local traditions.

Tompkins designed a series of fifty posters with a slogan he coined: "Let Corrientes Become Corrientes Again." The posters featured native animals that needed protection, including the maned wolf, the caiman, and the jaguar. The posters deliberately excluded the Tompkins logo or any indication of a foreign hand. Distributed free to kiosks, supermarkets, post offices, hotels, and bars, the posters proved to be a hit

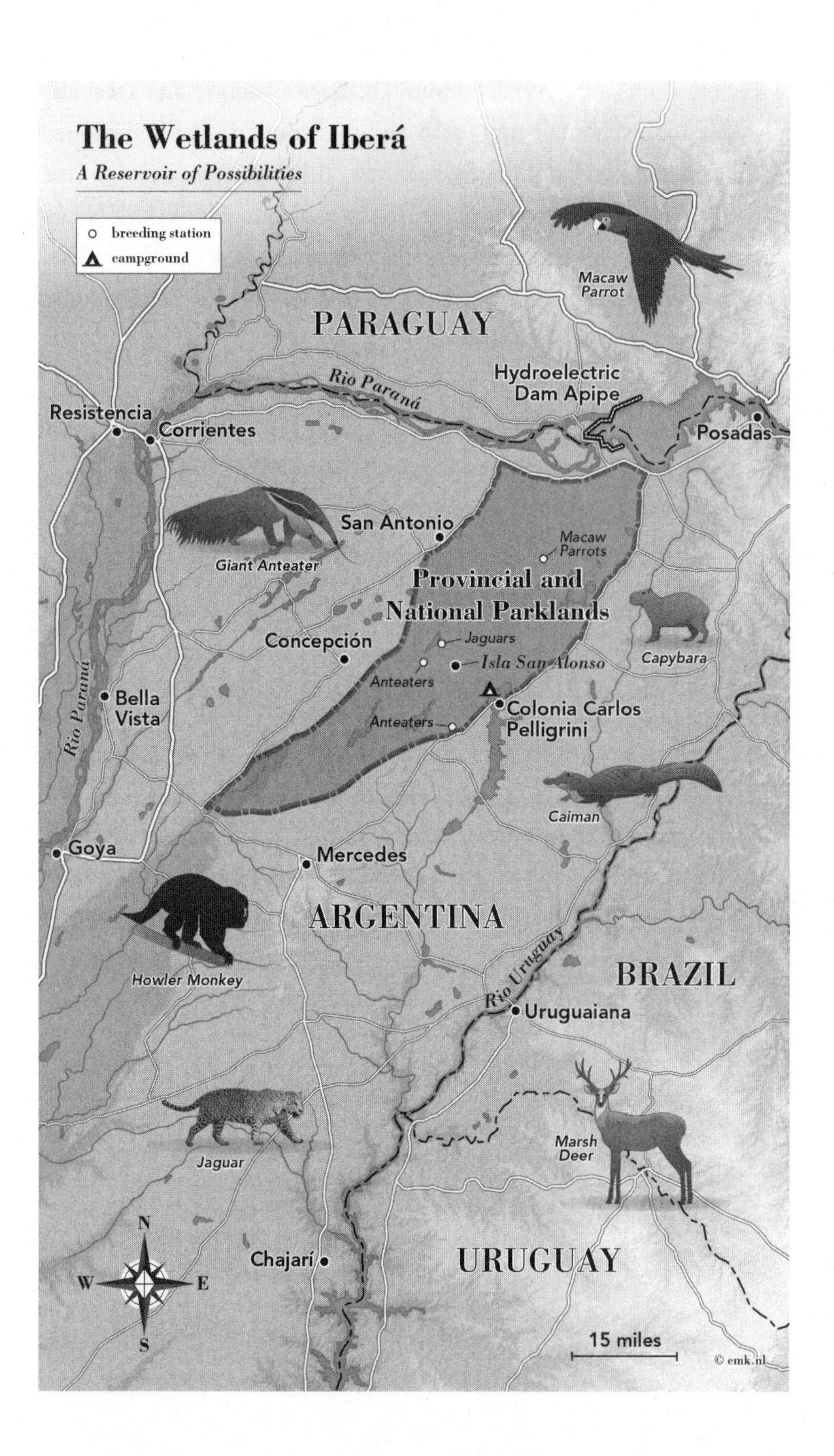
The Wetlands of Iberá
A Reservoir of Possibilities
breeding station
campground
PARAGUAY
Macaw Parrot
Rio Paraná
Hydroelectric Dam Apipe
Resistencia
Corrientes
Posadas
San Antonio
Macaw Parrots
Giant Anteater
Provincial and National Parklands
Jaguars
Concepción
Isla San Alonso
Capybara
Anteaters
Bella Vista
Rio Paraná
Colonia Carlos Pelligrini
Anteaters
Caiman
Goya
Mercedes
ARGENTINA
Rio Uruguay
BRAZIL
Howler Monkey
Uruguaiana
Marsh Deer
Jaguar
N
W
E
S
Chajarí
URUGUAY
15 miles
© emk.nl

with both adults and schoolchildren. The Conservation Land Trust also provided local school districts with study aides and classroom materials about the animals. Semester by semester, class by class, year by year, animals long extinct crept back into the marshlands and the school curriculum. Schoolchildren were learning to love the land and its animals. Given another ten years, Doug and Kris bet that the local connection to wildlife would grow significantly.

> *Doug was like Atlas, taking the weight of the world on his shoulders. You could see that, and it was definitely affecting his mood. He would get angry and frustrated. But what was so wild was how much beauty was in his life. He was dark, and there was gloom, it was kind of like doom porn that he would send out in these emails, another article from another dark, unknown writer from northern Europe. And I'd read it and just want to shoot myself because it was so dark. The gloom is coming, and there is no way out. So he was consuming a lot of dark stuff, but then somehow in his work, he was translating that into a solution, or an option, which was the creation of these beautiful farms, and these beautiful parks, the creation of these beautiful books, beautiful images. On one side what you had was doom and the end of the world, and on the other side he was cranking out beauty.*
>
> —WESTON BOYLES

The couple were sure that community support was essential. Otherwise their planned park in the Iberá wetlands wouldn't last a century. To further solidify their efforts, they also planned an economic renaissance for Iberá based on what they called "The Production of Nature," which argued that conservation opened a path to sustainable economic development.

For years suspicions toward outsiders and abandonment by Argentina's federal government had obscured the larger truth: Iberá was a wildlife and tourist gold mine. Kayaking, bird watching, nature tours, and African-style safaris could all become a base for a sustainable local economy. "What we offered was a paradigm shift, trying to give economic opportunity so that the ten towns had the chance to develop via wildlife tourism," said Heinonen. "The ten communities understood that we were offering something productive—the production of nature, via nature itself. This idea came about from seeing how the people in the area lived, because Corrientes is very poor. Trekking and hiking? They don't do that, so the local people can't connect to nature via recreational activities on the ground, but they *can* connect with it via work. That's why the idea of generating something that could give work opportunities, to produce something, made all the difference."

To convince local critics, Tompkins proposed taking local authorities on a tour of national parks in Africa. Sergio Flinta, the ever-less-recalcitrant Argentine senator, agreed to go. "My trip to Africa helped me to understand what it is to generate an economy based on wildlife," he said. "Seeing Kruger National Park was very important for me, because it is a mixed reserve, provincial and national. I got all that conceptual framework and applied it to our initiative. Iberá Park in Corrientes has a lot of South Africa in it, and a lot of Costa Rica. The African component is the reintroduction of species, and the Costa Rican component is the relations with local people and local development."

Locals had long doubted that tourists would fly to Argentina and pay to see an anteater or caiman. But as the stream of tourists grew, and more and more visitors rented boats, took photo safaris, bought meals and supplies, and stayed in local hostels, the lesson was clear: people *would* pay to see caiman and capybara. "Tourism is the easiest way to link fierce beasts like jaguars with jobs because people understand it," said Jimenez, the biologist. "Once they see other areas where people have improved their quality of life through ecotourism based on wildlife, local people say, 'Okay, we're going to do it!'"

For Doug, public access was also a crucial part of his strategy to protect the wetlands. He was certain that if locals began to see the wetlands as a magnet for tourists, they would consider it a wealth held in common. Then they would fight to protect the wildlife as they held a stake in the overall ecological health of the wetlands. Building a public dock and promoting the idea of exploring the inaccessible wetlands was an evolution from his strategy—a decade earlier in Pumalín—to oppose all development. Now he was sending his emissaries first, having them stake out the unclaimed land and teach locals to take a stand for nature by developing tourism.

Among those who noticed what Doug and Kris Tompkins were doing in Corrientes was Argentina's tourism ministry. These officials understood the value of developing Iberá with an emphasis on sustainable tourism as its development model. Not long after they grasped the potential of Iberá, they shuffled the country's tourist and infrastructure budget to pour millions of dollars into access roads, campgrounds, park guards, and marketing campaigns. Senator Flinta became a prized ally, not only for politicians, but for local businessmen and power brokers who suddenly realized that Tompkins was going to win the cultural war, as he had so audaciously predicted during his first talks at village meetings a decade earlier.

As the Iberá wetlands became ever more defined as an eco-tourism destination, they gained provincial acceptance, and then national and international recognition. A peculiar and noteworthy tailwind developed: everyone wanted to be part of it. "Those who were enemies and blocked access to the park now donate a strip of land for a road," said Flinta, who included himself among the converts. "Those who were staunch critics to public access now can't say a thing."

Chapter 19

The Route of Parks

Take the chain of seventy-seven national parks in the United States; not one of them was formed without a conflict or a polemic or something. You have to work through this. Some of them took sixty years to form. We did, I have to say, pretty well, in comparison with most of the American national parks.

—DOUG TOMPKINS

In 2013, as Tompkins approached his seventieth birthday, he panicked. He could still hike for six hours in the backcountry, and his ninety-six-year old mother, Faith, was healthy, but he worried his time was nearly up. "I have so many things I'd like to do before my hourglass runs out," he confessed in a personal letter. "Although in my heart of hearts I know nothing will stop the apocalypse, something in the system, almost genetically, propels me to work for beauty."

Tompkins was convinced of the need to save ever-more-vast tracts of wilderness. With the human population passing seven billion, he saw the window of opportunity to purchase intact ecosystems and to create new national parks slamming shut. "It was mind-boggling because he would be scheduled with meetings starting at eight in the morning all the way through the day, through dinner and sometimes afterwards," remarked Linde Waidhofer, the landscape photographer who worked with Tompkins. The donors, when they flew down, were provided tours and updates on the varied conservation projects, further adding to the busy agenda. A yearslong campaign to create Yendegaia National Park finally saved a key swath of Tierra del Fuego, and although Pumalín was still far from being accepted, the momentum was shifting.

Despite the clear evidence of a literally groundbreaking conservation project, money was tight. Generous grants from Patagonia Inc. and its founders, Malinda and Yvon Chouinard, filled key funding gaps. Yet funding was a perpetual struggle. "We get questions all the time from wealthy people interested in private conservation, asking 'How did you do this? Can we meet? How can I do something like that? How can I protect this area I love?' Doug always gave those people the red carpet treatment," said Carolina Morgado. "When people like that come into our world, they find it wonderful, 'All that Doug has done!' and lately, 'All that Kris has done.' While they are here with us, everything is possible, but one step out of this house, or one step out of the park, and the neo-liberal demand to produce, to do things that generate money, takes over. It's very hard to take that out of people. It's like a chip inside."

Despite a combined fortune measured in the tens of millions of dollars, cash was a limiting factor for Doug and Kris's conservation plans. Their conservation budget included salaries for a hundred employees in Chile and dozens more in Argentina. Rather than recruit a million individuals to donate a dollar each, Kris and Doug went for individuals who could drop a million. Doug taunted Nicolas Ibañez, who founded Walmart's Chilean operations, thus making him a billionaire. "So, Nicolas, when you die, what is your tombstone going to read—'Most important supermarket owner in Chile'?" Rather than take offense, the billionaire took notice.

The few dedicated "high-net-worth individuals" who made the long journey to South America for facetime with Doug and Kris were not disappointed. Evening dinner with the couple was an elaborately casual affair. "There were guests at their house almost every night," said the environmental activist and naturalist George Wuerthner. "Doug regularly met with political leaders, scientists, philanthropists, conservation activists, and well-known writers and artists. Sometimes world-class climbers or adventurers joined the group. There would be a spirited discussion."

Having scaled the summits of capitalism and conservation, Tompkins felt qualified to lecture his guests, even if they were celebrities at the level of the author David Quammen, or CNN's founder, Ted Turner. "Unless we learn to share the Earth with all the other creatures on the planet, our own days are numbered," he declared. "We need to teach our children that each person must pay his or her rent for living on the planet, and that means demanding of our governments to make biodiversity conservation a priority."

Emboldened by their success with Monte Léon National Park in Argentina and Corcovado National Park in Chile, Doug and Kris moved forward to consolidate a groundbreaking plan. What if they bundled all their Chilean properties into a single take-it-or-leave-it offer to the government? What if they raised the stakes, by putting all their land on the table? Doug knew this proposal was bold, and it would be debated by officials at the highest levels of the Chilean government.

I flew with him three times over Pumalín Park. On the last day, there was a lot of wind. I told him it was too dangerous to fly and he said, "Take it easy, this is like kayaking a river; you have to ride very carefully." We took off, and he told me to strap myself in well, because we were going to be bouncing around. I put my camera, a big Hasselblad, into a backpack. I had another small bag, with a Nikon FM2, a classic, and away we went. It was impossible to take photos while up there, as we were moving around like the inside of a blender. We made it to a part of Comau Fjord, which runs south to north. There's a section where you arrive to a fjord that's got a perpendicular form, Cahuelmo Fjord. When we got to Cahuelmo, a type of wind tunnel formed with the wind coming from the east. It all gets funneled together and blows really fast. That wind hit the plane, and the plane started jumping. It was really bumping around. Doug and I weren't talking. We had been wearing headphones, but his had fallen off because of all the movement of the plane, so we couldn't talk any more. I thought we were going to crash. When we landed, I took my Hasselblad camera out of the plane, then looked for my Nikon, and suddenly I saw that the transparent acrylic roof of the plane had a hole in it. The hole was in the shape of a camera.

—PABLO VALENZUELA,
Chilean landscape photographer

In 2014, the Chilean government inaugurated a ferry route through a maze of coastal islands in southern Patagonia. The southern highway ended in the bay of Tortel, a fishing village where travelers could now drive aboard a ferry and be hauled through a spectacular maze of islands south where the road picked up again in Puerto Natales, the gateway to Patagonia's most celebrated (and overvisited) national park—Torres del Paine.

This land-sea-land route passed through the most untouched terrain in all of Patagonia. On a map the region looked like a giant with a hammer had shattered the land into a hundred jagged pieces. No overland road was feasible, so navigating the archipelago by sea was the only way to connect the regions. When Tompkins understood that this new sea route would unite disparate segments and isolated national parks scattered along the Southern Highway, he had a wild idea. Perhaps his wildest.

Tompkins shelved his earlier arguments that the Southern Highway was an unpardonable scar on the landscape. Now he reimagined the rugged road as a spinal cord. He realized that the land and sea routes of the Southern Highway could connect a dozen national parks in Patagonia, from Alerce Andino park near Puerto Montt in the north, all the rugged thousand miles south to Tierra del Fuego. Tompkins needed an image, a unified concept. Then it came to him: he would rebrand the disparate ecosystems as a single entity.

Tompkins ordered his team to work crafting the campaign. He summoned a graphic designer, wildlife biologists, and key aides. Working morning to night, they designed a presentation for Chilean president Sebastian Piñera. Doug turned the concept over and over in his head, and was pleased. Here was a brand that would survive the test of time. After The North Face and Esprit de Corp, this third brand was designed to outlive the fads of fashion. For a lifelong collector, a man who cherished finely woven quilts and perfectly designed wooden chairs, this was his opus. He called it the Route of Parks.

Tompkins understood that the Route of Parks would encompass glacier-covered Andean peaks, coastal lagoons, and thickly wooded jungles draped in moss and decorated with forests of ferns. It was a flourishing biodiversity reminiscent of a more pristine era. Even if most travelers would visit no more than several parks, they had an exceptional menu from which to pick and choose. Doug had invented a buffet-style offer to tempt the world. "Doug was not interested in tourism at all," admitted Mladinic, who worked with Tompkins as an interlocutor with Chilean authorities. "Of course, we had lodges and trails because he believed that every park needs a basic public access infrastructure, but not because we wanted to have thousands of tourists coming. From his experience in Argentina, he was finding out that there was a healthy and fruitful partnership between tourism and conservation."

Using their contacts worldwide, Doug and Kris now pitched their conservation allies worldwide. This could be the couple's final push as they bundled their land holdings into a bold conservation megadonation. They would donate all their Chilean lands to CONAF—the underbudgeted national park service—and in exchange the Chilean government would create five new national parks, and enlarge three existing parks.

Tompkins put his design team to work. They prepared a highlight reel—a visual teaser—that showcased the benefits to Chile by branding Patagonia as a tourist mecca. Using stunning photos and simple slogans, he emphasized the benefits to Chile's national image. What could be more modern than a country announcing new national parks and sustainable tourism development plans?

Doug was determined to forever change Patagonia as a region and as a concept. When he met government ministers and even the president, Tompkins usually knew the map better than anyone in the room. He liked to run his finger down the spine of Chile, as he rattled off the parks, including new ones he thought ought to be created. Doug offered to donate roughly a million acres of his land if the government reclassified ten million acres as national parks. Leveraging their lands ten to one was a bold proposal. Even if it was only half successful, it meant the creation of millions of acres of new parklands.

Always an audacious negotiator, Tompkins boasted the street credibility of a pioneering outdoorsman and the intellectual vigor of a debate champion. Like the conservation leaders Iain Douglas Hamilton, Michael Faye, and Jane Goodall, he was among the few who lived amid the lands and animals he so passionately defended. "If you want to spend half a million dollars buying land, you will never be able to get more bang for your buck than giving that money to Doug and saying, 'Go for it!'" declared Buckley. "He can call the president, and talk to the cowboys."

Despite numerous advances with the Piñera administration, the mega-donation was never sealed. So, in 2014, with Piñera's term nearly over and Chile's national elections on the horizon, it was time to court the candidate most likely to triumph: Michelle Bachelet. The divorced, non-Catholic, single mom was not exactly the Chilean elite's ideal candidate, but she had already been president once for four years, and left office with approval ratings bordering 80 percent. In her first term, Bachelet had not connected with Tompkins. In fact, they had never met. With environmental protection a rising concern for Chilean citizens, however, making parks and saving wilderness increasingly garnered political prestige.

Tompkins had instructed his team to make a chart showing how many acres of national parks each Chilean president had created, a nature ranking. Since Chile's first national park in 1926 every president to complete a full term had added national parklands. It was a republican tradition; even General Pinochet had created national parks. Understanding the competitive nature of presidential egos, Tompkins created a chart that showed Bachelet as ranking second among all Chilean presidents for most acres of national parklands created. All she had to do was approve the Tompkins mega-donation plans. "We'll close the deal with this," Tompkins confided to Ingrid Espinoza, a trusted member of his inner circle.

"We emphasized the importance of local economies and how tourism was a consequence of good conservation," recalled Espinoza of the paper they crafted. "And that's how the new political proposal came about. More than park donations, we were talking about how the territory was going to develop, and we were contributors to that development."

He was showing us around the future Patagonia National Park and he was saying, "Well, yeah. And then when we give it to the government—" And I was thinking, How are you going to let go of this? How are you going to give this away? I mean, you're not even finished yet. *And he was such a control freak, and he did things at such a level that was completely unsustainable. Utterly. And he's saying, "You have to. You just have to. The ultimate goal has to be a national park, and you just* have *to be able to let go." I remember walking up the path with him by the greenhouses when he was saying that and I was, thinking,* This does not sound like him.

—QUINCEY TOMPKINS

In October 2015 Doug invited Fay to pay a visit and explore Patagonia. Fay arrived, and Tompkins took him up in the Husky and then banked the plane hard over, on its side. The wings tilted 45 degrees off the horizon. Doug spiraled lower, one wingtip toward the dark canyon and the other skyward. Around and around he corkscrewed into the crevice, wingtips whooshing just feet from rock walls. Fay—strapped into the seat directly behind Tompkins—wondered if they could get out. After agonizingly slow minutes of descent, the cavern opened, exposing a stunning waterfall surrounded by jungle, like a movie poster for *Jurassic Park*. Tompkins grinned. Fay smiled and thought, *Only Doug!*

Later, Doug picked up the rock climber Alex Honnold, the free climber who later scaled Yosemite's El Capitan in the Oscar-winning movie *Free Solo*, and with Fay provided a tour of hidden valleys, never-climbed rock faces, and smoking volcanoes. Tompkins beamed as he narrated the twenty-five-year history of his purchases, his properties, the parks, the fights, the farms, and the failures. "Flying with Doug over Patagonia is like being on a tour bus," explained Fay. "The conversation is constant, and it's mostly Doug showing you things. He flew me down to the salmon cages and we were counting them and then he goes through a pass and shows me the volcanic eruption and the forest that it destroyed and then he flies over another farm that he's fixing up and he shows me two or three buildings that they just rebuilt, and then he shows me the erosion from the volcanic ash in the river and how it's changed the channel. He's doing his tour, but at the same time he's giving you a narrative of what you're seeing. And at the same time, he's cataloging those things in his brain."

As he described the forests, the farms, the trails, and the views, Doug passionately outlined the grand scope of the Route of Parks. Like the art collection he had liquidated decades earlier, this collection was also worth a fortune. Doug never ceased in his proclamations that his goal was to give away *all* the lands, yet negotiations were frustratingly bureaucratic. Doug was eager to donate the parks, but President Bachelet's team was mired in a political crisis involving Sebastian Dávalos, the president's son. Her approval ratings plunged from the high seventies to the low thirties, as she fought for her political survival. Conservation was not a front-burner issue for her government. Doug's long-stated promise to bequeath all his lands to the people of Chile was yet again stalled. Purchasing the lands had entailed years of frustration, setbacks, and agonizingly slow progress. Now he faced the same challenges in giving it all away.

As work on the mega-donation slowly proceeded, Tompkins poured his energy into the design of a visitor's center at the center of the proposed Patagonia National Park. The center showcased his arguments for the creation of national parks, and the protection of nature. Jürgen Friedrich, his former Esprit partner in Europe, donated six million euros to cover construction costs for the visitor center. Tompkins envisioned the exhibits not as a rehash of the valley's history or a summary on the number of bird species but as a call to action. How best to shape the half hour that a typical visitor might spend inside the welcome center? "He wanted people to understand the notion of a national park very deeply," said Kris. "Why do you have to have national parks? He wanted to make sure they understood what is going on. That national parks are a necessity."

For Doug, the messaging was clear: he wanted to first shock the visitors, then fill them with facts, and finally motivate each visitor to take action. At the end of the experience, he planned an oversized mirror with the message "What Will You Do?"

In December 2015, while the visitor center was still under construction, Doug gave a tour of the half-finished exhibits to a group of business leaders who had come down with Rick Ridgeway and Jib Ellison on a program known as the Corporate Eco Forum. The executives worked at Fortune 500 companies, including Hewlett-Packard and Disney. They were exactly the decision makers whom Doug regularly savaged. Jib was more diplomatic. He saw a chance to teach rather than scold. In his mind, the largest corporations were often aware of their environmental sins and understood the urgency to reform their ways. Working with Walmart, Jib had convinced its owners, the Walton family, to reduce the company's packaging by 35 percent. That single accomplishment meant that Ellison's consultancy group, known as Blu Skye, had helped reduce landfill waste by thousands of tons.

Standing before the corporate leaders, Tompkins gave a virtuoso performance. He passionately laid out the arguments for a more profound commitment to conservation. As he spoke, his longtime friend Ridgeway could hardly believe what he heard. "He's walking through all the rooms even though there are no displays or anything, but he's telling them what it's going to look like," marveled Ridgeway. "I was listening to him describe this, and he started using pronouns that I'd never heard him use before, or never in that way. They were the pronouns of collaboration. They were *we* and *us*, and *our* problems, and *our* challenges, and *we* can do this together. I'd never heard him talk like that. It actually was different enough that I made note of it. I had a little smile to myself and said, 'You know what? Doug's been changing since he married Kris.'"

Chapter 20

Ambushed in Patagonia

All day long he was on the computer, which is kind of a crazy thing. He was talking with activists around the world, dealing with land issues. There was this great weight. He had this vision of land preservation and that time was running out, not only for the Earth. He really talked about that and kept saying, "I only have so much time to get this stuff done."

—EDGAR BOYLES, pilot, photographer, and lifelong friend of Douglas Tompkins

Doug Tompkins was grumpy and overworked. Stuck fourteen hours a day in the office, seven days a week, as spring sprouted around his remote home in Patagonia, he deserved an expedition to replenish his seventy-two-year-old body and soul, his friends suggested. It was time for another Do Boys adventure. Their idea was to traverse the remote north shore of Lake Carrera in sea kayaks over the course of five leisurely days. The word went out to the various Do Boy members, and Jib Ellison, Rick Ridgeway, and Yvon Chouinard all confirmed. With Doug that made four. Weston Boyles, at twenty-nine, sprang at the chance to join the elite group and, together with Lorenzo Alvarez, a highly regarded guide, added youthful strength and veteran safety to the journey.

Collectively, the six men held at least a hundred years of kayaking and rafting experience. Alvarez had competed with the US national rafting team, Boyles had navigated Class V rapids, and Ellison had guided rafting trips in Siberia. Ridgeway could simultaneously kayak rapids and hold a camera steady. During his twenty-year career, he produced adventure documentaries for ESPN, *National Geographic*, and the Patagonia company, where he held the rank of vice president. Although not the strongest paddler, Chouinard was unfazed by the ten-miles-a-day paddle, especially since they would be hugging the shoreline. Chouinard felt more worried about which fishing flies might work. Unofficially, he was lead fisherman. Tompkins described the trip as a chance to visit with friends and told a reporter he looked forward to "stretching the muscles a little."

The six-man expedition was divided into four kayaks—two doubles and two singles. Tompkins traveled in the same kayak as Ridgeway, while Ellison shared a kayak with Chouinard. Boyles and Alvarez paddled single kayaks weighted down with vegetables, boxes of oatmeal, bread, and boxes of red wine. The itinerary was relaxed: two days paddling, a rest day, two final days in the water, and arrival at Puerto Ibáñez on the eastern shore of the lake where their partners and friends would join them. Using the Google Earth app, they had scouted campsites along the route, and had seen interesting valleys for backcountry hikes. The lake's remote north shore was hemmed in by steep cliffs, and was empty of roads, fences, and practically any signs of human presence. Temperatures were forecast to fluctuate from the low forties at night to the high sixties during the day. At forty-six degrees south of the equator, below any landmass on the globe except New Zealand, sunset was not until 10:00 p.m. Evening campfires could be lit during daylight.

Chouinard flew in from California. Though they exchanged letters and occasional brief visits during overlaps of travel, this was his first expedition with his best friend since Doug's seventieth birthday bash two years earlier. Their relationship was difficult for outsiders to fathom.

They could visit one another for days and hardly talk. Fifty-eight years of friendship laid the foundation for deep understanding. Like the dynamics inside many a successful marriage, they understood one another without a need for many words. The competitive urges that once drove them as climbers and businessmen still burned. Doug always pressed to climb faster, paddle harder, but it was a jovial competition that also included secretly placing five-pound rocks in the bottom of one another's backpacks in hopes they wouldn't be discovered until the end of a strenuous climb.

On the evening before the trip, the six expedition members, together with spouses and a small group of friends, met for dinner at El Mirador de Guadal, on the shore of Lake Carrera. December in Patagonia was an ideal month to travel, with sunny days and star-studded nights. The high season for tourism was still a month away.

Lorenzo Alvarez looked down the dinner table and listened carefully. These were legends of kayaking and climbing. Besides first descents on half a dozen Chilean rivers, Tompkins also had bragging rights to a first descent in a kayak of the Zambezi River in Zambia as he skirted ten-foot-long crocodiles congregating just below the waterfalls, waiting for prey. Rick Ridgeway, sitting across the table, was Doug's companion on a half-dozen epic trips. He'd reached the peak of K2 in the Himalayas without using supplemental oxygen.

Tompkins and Chouinard laughed at their own frailty. Reminiscing about the romantic energy of themselves as young surfers, now they jokingly called themselves the "Done Boys," which quickly morphed into the "Never Done Boys." "The fact that we were going paddling for four or five days and trying to cover one hundred kilometers from port to port didn't even come up in the dinner conversation," Alvarez said later. "It wasn't like, 'Okay, let's pull out the maps and see how far we're going to go each day. And let's make sure we keep an eye on the weather.' It was all more like, 'Hey, we're having this great opportunity to see each other again!'"

We started the trip with a full-tilt ceremony at the Victoria Falls overlook, with the president of Zambia, and da-da-da. *We take off down this river. Doug is one of two kayakers; the rest of us are in rafts. I'm working from a raft, doing the filming. We're right at the base of Victoria Falls; it's just insane. Here's this wonder of the world, and you've got kayaks playing around near the base. It was an out-of-this-world vision. We head down the river, and go through these rapids that just drop into a placid pool. And boy, this is wonderful! Our Sobek guide friends said, "If a crocodile comes up out of the water, you can throw a rock at them and they go away."*

The crocodiles would sit in the pool below the rapids waiting for big fish to come through which they would grab. One day one of them grabbed one of the rafts, and exploded a section of the raft. There was a whole scene of Jim Slade from Sobek beating the crocodile over the head with a wooden oar to get him to let go of the raft. Doug is kayaking that and rolling. He was fearless.

We would sleep on the sand at night. It was really lovely because we didn't have to be in tents. There were no bugs. It was kind of like being on a desert trip in Utah. One morning we woke up, we've rolled up our sleeping bags and stuff, and we find these big tracks in the sand not very far from us. We realized that it was a croc that had either slithered up out of the water or had come down from somewhere back into the water, and went right by us sleeping in sleeping bags on the sand.

—EDGAR BOYLES

One of the few details discussed during the three-hour meal was the need to dress in foul-weather gear the following dawn. Winds whipping off the nearby ice fields would be bone chilling. Locals joked that visitors to Patagonia could experience summer, autumn, winter, and spring on the same day. "My very good kayaking buddies told me, 'You've got to be very careful of Lake Carrera,'" said Alvarez. "Because it's a glass lake but it can turn into a wild, wind-struck ocean."

The Last Voyage
Dec. 5-8, 2015
Puerto Sanchez
Río Muller
Camp 1
Camp 2
Río Avellano
Camp 3
Puerto Ingeniero Ibáñez
Day 1
Day 2
Day 3
Day 4
The Accident
Lago General Carrera
CHILE
ARGENTINA
Terra Luna
N
E
S
W
Rescue Boats,
Chilean Navy Base
Chile Chico
Los Antiguos
© emk.nl

At 5:45 a.m. on December 5, 2015, the six kayakers climbed into a small boat, dressed in rain gear and prepared for the onslaught of a bouncy ride, splashing waves, and a wind-chill factor near freezing. Tompkins sported cotton trousers, loafers, and a button-down wool sweater. He topped off the outfit with a beret that looked like a golf cap. "We were like, 'Well? Where's your foul weather gear?'" said Alvarez, describing the Where's Waldo situation. "One of the wives said, 'One of you doesn't look like the rest of you.' And Tompkins looked around, like, 'Well, *what*?'"

During the commotion of the send-off, Kris slipped a satellite phone to Ellison. Doug was not to know. Hi-tech gadgetry annoyed him. Forrest Berkley, a Boston-based philanthropist dedicated to land conservation, had given the phone to Kris, and she had written emergency numbers on the inside cover. She included the local police, the Tompkins Conservation office, her top aide Carolina Morgado, and the local Coast Guard station with jurisdiction for Lake Carrera.

All kayak expeditions on the lake were expected to file an itinerary with the Coast Guard as well as to contact duty officers during a daily check-in call. The Do Boys never gave a thought to informing the authorities. "Like that zen painter who always leaves part of his painting unfinished," Chouinard said, "we always left room for disaster."

When Phillipe Reuter had booked a reservation for a party of six with kayaks to cross Lake Carrera in his charter boat, he didn't realize the guests included Chouinard and Tompkins, icons inside the world's small circles of conservation, kayaking, and climbing. Reuter was a member of that same elite club. He had summited Mount Everest, skied the world's seven highest volcanoes, and lived a life of adventure. Reuter was a neighbor living on the shore of Lake Carrera and running his own business, Terra Luna Lodge. Reuter held a deep respect for the frigid lake. "Put it this way," he said: "I have lived here fifteen years, crossed that lake nearly every day, and never gone for a swim in it. . . . And when it gets rough, it changes from one moment to another. The wind starts up in the valley, then comes down to the lake and blows little waves at first. You can be bathing with your baby on the lake's edge in absolute calm and, in the distance, you can see a white spot, like a tsunami. One second it is flat, and then you are in the waves. That's the way it happens every day."

Tompkins shivered in Reuter's boat. Halfway across the lake, he was wet, cold, and in need of help. Although it was a cloudless, sunny morning, the pilot idled the boat while Tompkins piled on foul-weather gear and spent the remainder of the crossing guarding precious body heat.

Arriving at Puerto Sanchez, on the north shore of Lake Carrera, the men loaded gear into the kayaks, strapped on spray skirts to keep water from flooding in, and paddled east along the lake shore. The horizon was framed by a wide blue sky, valleys, and a rocky cliff-studded shore. Except for an abandoned gold mine and a few farms there were no roads, no trails.

The pace was fast, even a bit competitive as the men raced across the glassy water. Swells the size of champagne glasses pushed them gently, while a warm sun shone overhead. Every hour the men brought their kayaks together. They held onto one another as they talked, drank water, shared granola bars, and drifted as a single raft.

Tompkins suffered in silence. The rudder on his kayak was broken, making steering more difficult. He had not kayaked much in the previous couple of years, and his elbow was grinding in pain. Tough and proud, he raked the lake with his unconventional technique—flipping the paddle backward and reversing the hydrodynamics that, he argued, gave him a better draw. The group paddled three miles past their initial campsite.

At the campsite the men prepared their tents as Yvon grilled food on the campfire. At dinner Jib Ellison described a new sport: kayak sailing. To get them flying across the lake, Jib had brought kayak sails, which looked like old-fashioned spinnakers. Puffing ahead of the kayak, the sail had a parachute-like shape that worked when running with a tailwind.

Rare was the day when a spinnaker was more useful than not, since the lines could tangle and were susceptible to shifting winds. But

Ellison and Boyles had charted the wind flow on the lake. On paper, it looked like a chance to paddle less and glide more. A constant tailwind could push them east, from campsite to campsite. "If you have prevailing winds on a regular basis, then the water builds wind swells. It's a time-tested methodology for getting from point A to point B," explained Ellison, who had bought the sails online and hauled them to Patagonia. "You can eat and talk, and you're just flying downwind. You're sailing and kayaking! That was my vision."

On the morning of the second day, as they left the makeshift campsite, kite-sailing conditions were near-perfect. As the kayakers paddled east along the north shore of Lake Carrera, a tailwind formed sloping waves that rolled them forward, adding an extra push, like an invisible extra paddler. The kayaks rode the two-foot waves easily. Ellison paddled and felt content—this was the way to travel, "like skiing," he remarked as the group coasted down little wave after little wave. Arriving in synchronized sets, each wave was a push, a chance to leave the paddle in the water and glide. "When the waves are twenty inches or less it's gravy," explained Ellison. "But when they get bigger and start breaking and becoming irregular in shape and size, then it's a whole other story. Then, very quickly, it goes from Class III to Class V."

As the rest of the group paddled forward, Boyles stopped to rig his sail, figuring he'd literally swoosh by the others within a few minutes of launching his sail. Aided by the hearty tailwind, the two double-kayaks sped ahead. Alvarez, paddling full force, was unable to keep up and fell further and further behind. There was no way a single kayaker could out-paddle the double.

Boyles fought to untangle the knot of a mess that was his kite kit. As he struggled to launch the sail, he was left far behind and disappeared from sight. Alvarez reacted with frustration. When the team reached shore, Alvarez scolded the Do Boys. From an expedition guide's point of view, they were reckless. "What if I had flipped? You guys never even looked back!" he exclaimed. "I lost track of Weston five minutes after we launched. And I don't know where he is now. He could be swimming, for all I know."

The five men climbed up a rock to get a better vantage point. Atop the hillock they sighted Boyles, paddling slowly, upright and apparently fine. They were all shaken by the incident. "You have to wear a damn life jacket," said Alvarez, who was indignant as he lectured Tompkins, a veteran outdoorsman he barely knew. Tompkins absorbed the attack, placid and confident. "If anyone falls into the water, they are gone," Alvarez pleaded. Tompkins knew the lecture: "Play by the rules and you will be fine." As Alvarez outlined the blatant dangers confronting them on the five-day expedition, Tompkins barely took his eyes off Lake Carrera, a lake twenty times the size of Manhattan with weather so intense that the Chilean Navy just days before had delineated areas in which kayaking was "strictly not recommended."

"There was a part of me that definitely didn't want to sound like the worrywart and pain-in-the-ass guy on the trip," said Alvarez. "But, being a commercial rafting outfitter, I always have safety be the basis for any fun." He suggested that the men use a buddy system "because if anybody tips over out here and there's nobody to assist you, you're pretty much screwed." Alvarez remembered Tompkins's lack of reaction to the warning, thinking that "his reputation or his personality was one where you couldn't tell him anything, because he knew it all."

The trip was notably and deliberately leaderless. Evening meals were a potluck communal affair. Someone brought beans, another chipped in red wine. Everyone carried a food bag, but there was little coordination or order to the provisions. Their bread had become sodden, so they laid out every piece on an old board to dry it out. For cookware, the Do Boys had brought a single pot with a small top they used as frying pan. Ellison forgot his spoon, so he found a piece of wood and carved one.

Tompkins and Chouinard never felt more comfortable than in moments like this. They had forged their friendship just a few hundred miles south in the ice caves of Mount Fitz Roy. In 1968, when they first arrived in Argentine Patagonia, Chalten was just a gaucho stop. The foothills were populated by feral horses and huge sheep ranches. Today it was an ecotourism hub with some 1,500 permanent residents. Over the last forty-plus years, both Tompkins and Chouinard had dedicated themselves to protecting the Patagonia region. And they had taken vastly different routes.

Yvon had named his company "Patagonia," brought his whole team to find inspiration in the vast region, and dedicated his business to "inspire and implement solutions to the environmental crisis." Through his initiative One Percent for the Planet, Yvon continued to raise millions for environmental programs around the world. His bestselling book, *Let My People Go Surfing*, had enshrined him as the world's coolest businessman. When he described himself, he often used the phrase "The Education of a Reluctant Businessman." For years he had also generously donated millions alongside Doug and Kris on joint conservation projects, including Patagonia National Park. While Malinda and Yvon were often low-key about their contributions, the impact was huge.

Doug was refusing to slow down, as he believed in pouring his every moment into conservation. Business had provided the wealth to fund his conservation dreams. Together, he and Yvon had done more than practically anyone else to define Patagonia as a wild landscape, and one in need of a collective defense. They had also seeded that message around the globe.

Now, arriving at their beachside campsite on Lake Carrera, they found their landing spot spoiled by a raft of spiny sticks and a dead cow. They piled away the sticks and hauled off the stinking cow, and then when the beach was clean and orderly, just as Doug liked, they set up their evening meal and unrolled sleeping bags.

Tompkins woke just past dawn, ate a bowl of oatmeal with dried fruit, and confided a secret to his pals. High above their camp spot on a plateau above the Avellano Valley was a property he wanted to purchase. It was so remote that a team of oxen would be required to pull tree stumps and clear the land for a short runway so he could land his Husky. He warned his friends "Don't tell Kris!" as she felt he had enough projects already, with all his other farms and their national park projects. Tompkins described the property in the valley as a secret hide-out, a "final place to come think."

This day of the trip was planned as a hiking day—no kayaking. Instead they could stretch their legs and explore. While they waited to leave, Tompkins climbed a nearby boulder. It was steep and, without ropes, a dangerous sixty or so feet up a rocky embankment to a knoll. Twenty feet off the ground, climbing in boat shoes and with no ropes, Tompkins froze. He was on a tiny ledge and needed to scoot over and up, a move that in his prime was second nature. The pause stretched for uncomfortably long seconds. His feet were balanced—"smearing," in climber jargon—on tiny footholds, and he couldn't move either hand. "He doesn't look like a climber," thought Chouinard. Without ropes, climbing down was impossible. A few seconds later, Tompkins slid along the ridge, found a route, and pulled himself to the top.

The men continued up the slope as a group, but in the rough terrain were soon separated, each bushwhacking and bouldering higher up the valley at his own pace. Ellison lost Tompkins. Was Doug left behind? Ellison continued hiking, slightly worried but figuring they'd bump into one another eventually, perhaps at the top of the valley. Climbing a ledge to scout for Doug, Ellison was surprised. Tompkins was not lagging behind but was nearly a mile ahead of the rest of them, his white cap bobbing through the underbrush as he sprang up the trail.

Weston Boyles caught up with Tompkins first. He found him sitting at a scenic overlook of the Avellano Valley. While they waited for the others, their conversation covered some of the finer points of the Route of Parks. They had spent much of the last year mapping every section of the 1,700-mile-long route. Tompkins flew at low altitude and Boyles traveled by van. Tompkins reviewed even the smallest details. What color did the Public Works Department paint the bridges? Was the widening of a new section of road aesthetically pleasing, or brusque?

When the rest of the group reached the valley overlook, they headed together toward the small farm that Doug was thinking of buying. Alvarez and Tompkins brushed off the earlier altercation as they walked and wandered deep into a conversation about organic honey. Tompkins explained how Pillan, his commercial honey operation in Reñihue, had grown in less than a decade to become among the world's largest producers of organic honey. Pillan had customers in Europe, Asia, and the United States. Alvarez was impressed. Tompkins commanded a detailed understanding of the minutiae of beekeeping, honey production, and even organic export pesticide-residue standards. Alvarez realized that Tompkins was not only a daring rock climber but also a nerd.

Bushwhacking all afternoon kindled their appetite. The evening dinner, regardless of Yvon's luck fishing in Lake Carrera, would be a feast. Hunger was always the best sauce. During their meal, the six men reflected on their beautiful surroundings. Tompkins described how, over the course of twenty-five years living with locals in Chile, he'd seen them increasingly incorporate the natural beauty of these lands as part of their identity. Patagonians, he said, now fought to protect many of the same waterfalls, mountain peaks, and wild rivers that first attracted him to the area.

After the conversation by the campfire wound down, Tompkins zipped into his sleeping bag and, in the lee of the rocks, slept protected from the increasingly harsh wind. The campsite was so sheltered that when violent gusts of winds roared overhead they barely touched the group's peaceful oasis.

Early the next morning, the Do Boys ate breakfast, packed their gear, and prepared for a full day of kayaking. Doug drank his customary hot water, then spent twenty minutes scouring the cooking pan with sand to leave it shiny—just the way he liked it.

December 8, 2015, was a national holiday in Chile as Catholics commemorated the miracle of Immaculate Conception. Across the nation, loyal churchgoers attended Mass and nearly all government offices were closed.

As they left their campsite, Weston Boyles launched first. He kayaked ahead of the group and turned frequently to snap photos of the five men following behind. Chouinard shared a kayak with Ellison. Tompkins paddled a double kayak with Ridgeway up front. Alvarez manned the other single. As the expedition pulled away from the shoreline, the wind churned small whitecaps on Lake Carrera. Two-foot waves rolled through in clean sets that felt challenging but not dangerous. The tailwind allowed them to surf down the waves, but conditions were getting worse by the minute. Their protected campsite had disguised an imminent danger: a storm was blowing in. Unbeknownst to the men, throughout Patagonia a weather alert had been posted. Harbors, airstrips, and docks were ordered shut. A gale was brewing, and the Coast Guard ordered ships back to port.

On Lake Carrera, the explorers debated briefly then decided to call it a day. The only question was: Where to land? Scouting the coastline, they pinpointed a hammerhead-shaped peninsula dead ahead, across a small bay. Instead of hugging the shore, they could navigate straight across the open stretch for perhaps a mile, and then find shelter in the lee of the rocky peninsula. "Within five seconds of having let go, we were not able to communicate anymore," said Alvarez, who described waves cresting so high he was unable to see more than a hundred feet in any direction. "Right away, I realized we were in way over our heads. And I was quite nervous for everybody."

Ellison and Chouinard in the double kayak went first, followed by Alvarez and Boyles. The group was blown apart. Their tightly bunched line of kayaks was separated by a length of several football fields. Tompkins and Ridgeway banked closer to shore, on a distinct heading, causing others to wonder: Had Doug and Rick decided to crash-land on the rocks, then haul their gear along the shore to the rendezvous spot?

An ice-cold wind poured down the Avellano Valley, where they had hiked the previous day. That wind collided with the westerly tailwind. The two air currents formed a dangerous whirl of waves. Tompkins and Ridgeway found it difficult to keep their kayak aligned with the uneven sets of waves. Unlike a nimble river kayak, sea kayaks are more sluggish. They can be turned, but it requires more strokes. And if a sea kayak is blown sideways to a cresting wave, it rolls. "It felt like the paddle was going to get ripped out of my hands," said Alvarez. "It wasn't just your ordinary Patagonia strong wind. It was a Hell-wind."

Alvarez ran safety checks in his head. Food—yes, they had enough. Schedule—fine, the day-to-day activities were flexible. Fire—they had lighters, fuel, and matches. Convinced they could wait out the storm, Alvarez focused on the kayak directly ahead of him and paddled with all his might. It was an exhausting, technically challenging route but he figured that at most it would take half an hour to reach land, unpack the kayaks, and settle in at their new campsite.

Behind Alvarez and closer to shore, Tompkins and Ridgeway lost their battle to stay atop the waves. Their double sea kayak flipped when they were roughly 100 yards from shore. After attempts to right the kayak failed, Ridgeway cinched tight his life jacket and they jettisoned the kayak.

Tompkins swam toward shore. The surf was rough and, together with Ridgeway, he might get bashed on the rocky beach, but they had each surfed enough to survive a rough landing. But as soon as they started to swim, a strong current pulled them in the opposite direction, away from the beach. "Doug and I had no way to know whether our companions in the other boats, who were ahead of us and out-of-sight, knew of our predicament," explained Ridgeway. "We realized we had thirty minutes, perhaps a little more, to survive." The crashing waves washed over Ridgeway and Tompkins. They were bitterly cold and starting to drown. Ridgeway assumed his life was over. "For a few minutes," he admitted, "I gave in—just let go."

Boyles, Alvarez, Ellison, and Chouinard arrived at the sheltered peninsula, and regrouped. Seeing no signs of Tompkins and Ridgeway, they climbed a bluff and nearly gagged at the sight. An upside-down, empty kayak floated and nearby two bodies bobbed in the water. They watched in shock as Doug and Rick attempted to reach shore. The waves were poised to toss them onto the beach, perhaps even thrash them on the rocks. Boyles calculated they were nearly close enough to reach a safety rope. He could toss a lifeline, pull them ashore, build a fire, and warm them up. "Oh, boy, this will be a shit show," thought Boyles, "but we'll be okay."

In less than a minute it was clear that Doug and Rick were being swept toward the center of the massive lake. The Do Boys scrambled. Ellison, Alvarez, and Boyles lightened the load in their kayaks, while Chouinard assisted in snapping on the spray skirt for Boyles, who paddled into the maelstrom. Chouinard kept lookout from atop the bluff. He never took his eyes off the bobbing figures. Alvarez and Ellison set off in the double kayak, which carried an extra paddle plus the smuggled-in satellite phone. Boyles led the charge in a single kayak.

Paddling out of the lee of the peninsula, the rescuers were punched in the face by the wind. They steered toward the last spot they'd seen their lost friends—who were drifting fifty yards apart. Chouinard scouted from his hilltop vantage point. He saw Tompkins swimming, actively fighting. Ridgeway looked dead, his inert body floating like a corpse.

Boyles tore into the water with his paddle as Ellison and Alvarez followed behind him. Boyles navigated into choppy waters that looked more like an ocean than a lake. The spray, the swells, and the wind made paddling difficult. Boyles stroked hard and followed the gestures from Chouinard, who was atop the rocky outcropping directing him toward Tompkins. "Doug had a dark, wild, catlike look in his eyes," said Boyles. "He was kicking as hard as he could while holding onto my boat." Gripping the tail, exhausted and half frozen, Tompkins sputtered orders. He took command of his own rescue.

Ellison and Alvarez found Ridgeway minutes later. He was floating on his back, unresponsive, and apparently unconscious. But when they yelled at him, Ridgeway groaned, came to life, and latched an arm onto the bow of their kayak. With Ridgeway on the bow, forward progress was impossible. Hand over hand Ridgeway crab-walked along the edge of the kayak. One false move and he'd be whipped away by the wind and currents. Ellison went into rescue mode. In a voice loud, slow, and clear, he described to Ridgeway a near-fatal accident they had survived during a rafting trip in Siberia. The nonstop cheerleading annoyed Alvarez, who worried that it drained energy better used to paddle, but the verbal motivations worked. Ridgeway aligned himself with the stern, allowing Ellison and Alvarez to swing around and paddle toward Boyles and Tompkins—they could help in that rescue.

A few strokes later, they realized it was futile. They too were being dragged away from shore. Instead of helping Boyles, they put all their attention toward saving Ridgeway—and themselves. Paddling with full concentration and a dose of adrenaline, they still could not advance. The horizon was unchanged, and they were no closer to land.

Ellison spotted a small outcropping and in a final push they managed to pull their kayak onto the rock and escape the wind. They propped Ridgeway in the front of the kayak while Ellison and Alvarez unpacked the satellite phone and called for help, but it was early morning on a national holiday and no one was picking up. Attempts to raise the Chilean Coast Guard went unanswered. Desperate to see if the phone even worked, Alvarez called his girlfriend in Truckee, California. Though it was 5:30 a.m., she picked up and patched through information to Carolina Morgado at the Tompkins Conservation offices in Puerto Varas. Morgado raised the alarm with a pilot, Rodrigo Noriega, who dialed his contact at Terra Luna Lodge—the same Philippe Reuter who had organized the transfer of the group across the lake three days earlier. Morgado also rang the Coast Guard, which dispatched a rescue boat from Chile Chico that, due to the wild weather, would take nearly an hour to arrive.

Then Ellison, with the shivering Ridgeway aboard, paddled for shore, where he made a fire. "I got Rick out of all his wet gear, and put him in a sleeping bag naked. I had a small fire going, as there was plenty of dry wood there. He was breathing, but still not responsive. As soon as he was in the sleeping bag—the fire was five feet away from him—I looked at him and said, 'I've got to go get Lorenzo. I'll be right back; hang in there.'"

Reuter was working outside when the emergency call arrived at Terra Luna Lodge. He ran to find Alejandro Maino, a veteran of the Chilean Air Force. Together they scrambled to launch an emergency rescue flight using their Eurocopter B3, a helicopter model so strong it once summited Everest. Configured for transport and tourism, the Terra Luna helicopter sported neither a winch system nor specialized rescue equipment. Reuter figured they would find someone on the edge of the lake, waiting to be evacuated and perhaps injured. He grabbed a life jacket, a climbing rope, and an orange life buoy ring similar to those used at public swimming pools. Within five minutes of the truncated phone call from Morgado, Reuter and Maino were airborne, speeding at seventy miles an hour across the white-capped lake, headed toward El Avellano Valley. "We were looking for people," said Reuter, who rigged a harness allowing him to lean far out the open door and scout the surface of the heaving lake.

Pulled farther and farther out toward the center of the lake, Boyles fruitlessly paddled as Tompkins attempted to haul himself aboard the back end of the single kayak. With less drag, Boyles might be able to overcome the wind, conquer the currents, and have the energy to paddle them back to shore. As Boyles turned in an attempt to yank Tompkins out of the water, his spray skirt unbuttoned. The kayak sank deeper. Boyles managed to tug his wool hat onto Tompkins, who clenched his waist with an iron grip. Tompkins kicked with all his force as together with Boyles they angled toward shore, Tompkins shouting out advice about where to hit land. "He knew he was fighting for his life," said

The Accident
10 am, Dec. 8, 2015
to the hospital
emergency bonfire
capsize
rocky island
Ridgeway rescued
helicopter rescues Tompkins
Tompkins drifts
water temperature 38° F
Lago General Carrera
N
W
E
S
Ridgeway
Tompkins
© emk.nl

Boyles. "He had so much more work to do. He wasn't going to give up and freeze."

"Doug was conscious with me for at least twenty to thirty minutes fighting toward shore, kicking, and I was paddling as hard as I could," said Boyles. "When he lost consciousness I had to keep his head out of the water," he explained. "When I turned around, I lost my paddle; it went away. I was cradling him, trying to paddle with my hand." For half an hour the two men drifted further and further into the center of the lake. They were now several miles from shore.

From his perch in the helicopter, Reuter spotted two figures on a beach but when the helicopter hovered and prepared to land the men on the ground yelled "No, No. *No*!" and signaled with their hands—gesturing toward the center of the lake. The helicopter rose and blasted back out into the lake. "Finally, we found a kayak," said Reuter. "Alone. Nobody in it." Flying another five minutes, Reuter spotted Boyles, sitting in a single kayak. He struggled to embrace Tompkins, whose body was listless and dragging in the water. The wind was propelling the drifting men even further from shelter, and Tompkins had been half submerged in the thirty-nine degree water for well over an hour.

Even from his vantage point inside the helicopter, Reuter kept losing sight of the men in the troughs of five-foot swells. Waves washed over the kayak, nearly ripping Tompkins away from Boyles. Lake Carrera was a clash of currents. Waves smashed and fused in a chaos that made piloting the helicopter risky business.

As he descended, Maino relied on his instrument panel and commands from Reuter, who shouted every few seconds, guiding the helicopter upwind. Reuter tied a carabiner to the end of a rope and lowered it to Boyles with care. If the rope whipped into the rear rotor, they could all die. Maino lowered the helicopter until it was just ten feet above the waves. The entire rope was only sixty feet long—half the standard issue used by the US Coast Guard. With winds gusting to 50 mph, the margin for error was minimal. Maino maneuvered the rope ever closer to Boyles. Finally, Boyles grabbed the lifeline, clipped the carabiner onto the deck of the kayak, and clutched Tompkins.

Gripping the safety buoy, Boyles was almost stretched apart. One arm, elbow locked into the rescue ring, kept them attached to the helicopter. His other arm braced the weight of the kayak, plus the drag from Tompkins, who was being pulled through the water. Rotor wash from the chopper blasted them both with gusts of freezing-cold air. Boyles's face contorted in pain. Reuters shouted for Maino to slow their advance just as Boyles indicated he needed the helicopter lower. Maino refused. The rotors were spinning so close to the waves he feared they'd catch a crest and tumble the helicopter into the lake.

At an agonizingly slow pace, the makeshift rescue operation dragged the kayak toward shore. Then in an instant, the kayak flipped. Boyles floated and Tompkins sunk below the surface. Reuters thought: "This is the beginning of the end, but surprisingly it was for the good."

Unable to regain control of the kayak, Boyles stuffed half his body into the rescue ring, rolled onto his back, and, using both arms, hoisted Tompkins atop his chest. Without the instability of the kayak, the operation moved far faster, yet it still took another thirty minutes of towing them through the freezing water before the helicopter neared the rocky cliff that was the shoreline.

Boyles let go of the safety line and swam them both to shore, maneuvering a crashing surf break. He laid Tompkins out on a tiny rocky ledge. Maino balanced the helicopter with one skid on a boulder while Reuter jumped off. The rotors spun less than three feet from the face of the cliff as Boyles shivered uncontrollably. "Don't worry! You're going to be okay!" Reuter shouted.

Boyles had spent nearly two hours harnessing every last reserve of energy. Sweating hot, then burning cold, his muscles twitching in agony, his numbed mind exploded in pain. His childhood hero was white-faced, sodden, and perhaps dead.

Despite repeated efforts to hoist Tompkins aboard the helicopter, it proved impossible. Boyles could barely stand, and Reuter weighed twenty pounds less than Tompkins. "The wind was gusting so hard that the helicopter kept buffeting wildly and it felt as if the rotors were going down and would cut us to pieces," said Boyles. With the helicopter precariously balancing on the boulder, Boyles climbed in and they flew along the lakeshore to assist the others. Tompkins was alone, on the beach and half frozen.

Landing a few miles away at the other rescue site, Reuter spoke: "I regret to tell you but Doug is dead!" Chouinard broke into tears. Weston refused to give up. He suggested that several of them fly back and try to load Tompkins, then fly him to the nearest hospital. Hypothermia experts often say a patient is not confirmed as deceased "until they are warm *and* dead." Extreme cold essentially puts certain bodily functions in hibernation, and Alvarez knew about hypothermia cases in which patients thought to be dead had recovered after far-worse exposure.

Ellison and Alvarez, the only two expedition members with any remaining strength, climbed into the helicopter. Back at the rocky beach, as Maino hovered above Tompkins, the two men jumped out of the moving helicopter. Working together, they hefted Tompkins into the bay of the chopper. Flight time to the Coyhaique Hospital was just fifteen minutes and, having been alerted to the hypothermic condition of the incoming patient, a trauma team awaited.

Hearing about the accident in Chacabuco Valley, Kris needed to be alone. "I left everyone in the office and moved like a stone out in front of the restaurant where both the Cessna 207 and Doug's Husky were tied down," she said. "I crawled under the belly of the Husky and lay flat against the grass, digging my fingernails into the dirt moaning a sound like wolves calling in the night. I would not leave the Husky and I held onto the grass with all my might until someone pulled my feet and drug me out. I knew that if I lost contact with his Husky, I would lose Doug."

Dagoberto Guzman, the Patagonia Park superintendent and a longtime collaborator of the Tompkinses in many conservation projects pulled Kris into a 4 × 4 and sped toward the Coyhaique Hospital. It was a dangerous, gravel road journey that could take at least six hours. But they found the road closed for repairs. Jumping from the car, Dago informed the road workers that "*Don* Doug is injured" and that Kris was in the car, rushing to see her husband. The workers opened the road and as Kris rode by, each man removed his helmet, placing it over his chest in silent respect.

At the Coyhaique Hospital a team of doctors and nurses battled to revive Tompkins, who had arrived with a body temperature of sixty-nine degrees. Witnesses told the medical team that Tompkins had been in the frigid water nearly two hours. As Kris sped closer, news spread of Doug's accident. A crowd gathered outside the hospital. Word ricocheted throughout the close-knit Patagonia community that Doug Tompkins had been gravely injured. Then news arrived that Doug was dead. But soon after, the news was uplifting—Doug was alive again. He'd been revived! The messages were incomplete, contradictory, and terrifying.

Kris followed it all via phone and text as she and Dago careened toward the hospital. Mladinic, the sociologist and key aide, was inside the emergency room. He provided minute-by-minute updates, allowing Kris to hear firsthand the state of her husband. As Tompkins slipped away, Mladinic placed the phone by Doug's ear and told Kris it was time to say goodbye. Fifteen minutes before she arrived, just after 6:00 p.m., the unbelievable news was confirmed: Doug Tompkins was dead. Kris collapsed. Her Eagle was gone.

Chapter 21

Year 1 A.D.—After Doug

If your life's work can be accomplished in one lifetime you're not thinking big enough.

—WES JACKSON, founder,
The Land Institute

Yvon Chouinard was in shock. In the lee of the blasting wind, the veteran outdoorsman sat in tears. Rick Ridgeway lay naked in a sleeping bag, shaking in convulsive fits. Weston Boyles, bundled in the same sleeping bag as Ridgeway, was incoherent, his lanky 6'3" frame unable to hold heat, his brain on fire, and his muscles exploding in pain. After grappling with waves, wind, and the weight of towing "Uncle Doug" back to shore, Boyles felt broken. Around a campfire a few yards from the crashing surf of Lake Carrera, the remaining three members of the six-man expedition huddled as they waited for the Chilean Coast Guard rescue ship.

The Coast Guard bundled the three men into the cabin of their patrol boat and motored back to port in the village of Chile Chico. The officials needed to debrief the survivors. There were many questions, and the media was barraging them with frenzied calls about the fate of Doug Tompkins. There was paperwork to be filled out, questions to be answered, and an overabundance of Chilean bureaucratic protocol. To the shell-shocked survivors, the interrogations felt like torture. They desperately wanted to abandon ship and go home but the formal investigation stretched on hour after hour. Finally, they were permitted to leave. Chouinard, Boyles, and Ridgeway boarded the Eurocopter to fly back to Terra Luna Lodge where, four days earlier, under a perfect dawn sunrise, they'd chortled and laughed while embarking, like schoolboys playing hooky.

Flying into the wind, fighting gusts, pilot Alejandro Maino sweated as he maneuvered his helicopter. The Eurocopter engine boasted 850 horsepower and a top speed of 160 mph, yet Chouinard sitting in the back couldn't believe it—the chopper didn't advance. Were they also doomed? Would they have to abort the flight, turn around, fly downwind in search of a clearing for an emergency landing? To no one in particular, Chouinard growled that it was the "slowest fucking helicopter flight of my life."

Maino's experience flying military aircraft (especially under simulated battlefield conditions) guided the craft home. After twenty minutes of jarring turbulence, they hovered near Terra Luna Lodge, then touched down. The survivors were welcomed with hugs and wettened by tears. How? Why? Where? All that could wait. Raw emotions poured out as the lodge filled up.

In a villa outside Milan, Italy, as the photographer Oliviero Toscani brewed morning coffee, he was jolted when the news of Doug's accident rattled over the airwaves. Doug Tompkins! His co-conspirator! Dead? Toscani was irate. He wanted to smash the coffee pot. For all those years that his friend Doug had invested in nature, protecting and cherishing wilderness, hadn't he been screaming "Be careful! Nature is a brutal beast"? Oliviero had scolded Doug how many times? Toscani felt the weight of his warnings. "I'd always told Doug nature is dangerous. 'Why do you work so hard to be so close to nature?' And then, nature killed my friend. I hate nature."

Mike Fay, the *National Geographic* explorer who flew often with Doug, was in Tanzania when "Conservationist Dead" scrolled on the screen. Immediately he knew it was his friend, ally, and copilot. "Conservation is kind of like war for me," said Fay, who had mourned dozens of colleagues, some murdered by poachers and others by anonymous assassins. "When something like this happens, you're not surprised at all. You're not shocked. You're not in disbelief. You are ready for it already. It's not necessarily a period of sadness because you've already absorbed that sadness before it happened."

In London, Sir Norman Foster was "beyond disbelief" when he heard that Tompkins had died in a kayak. "It was as if they were physically and mentally joined," said Lord Foster, the architect who designed Esprit showrooms with Tompkins in the '80s. "Doug's energy was contagious," he mused. "He had a combination of intensity, curiosity, and conviction. Normally somebody with conviction will have a closed mind, but I didn't sense he had a closed mind; I sensed he was open. He was a force of nature."

In Argentina, senator Sergio Flinta did not know that Tompkins was drowning when he entered the presidential palace in Buenos Aires. Flinta presented conservation plans to the tourism minister, Gustavo Santos. "I brought the Tompkins Conservation books, to tell him about the Iberá project," said Flinta. "And Santos told me, 'This is an extraordinary project—marvelous!' and took it as his own, which was perfect. When I got back to my hotel, I got a call telling me that Doug had died. At the same time that he was dying I was getting the [Argentine] national government to take the Iberá project as its own."

Kris was distraught, living in a "well of ache." She and Doug had spent numerous nights formulating contingency plans for the accidental death of one, the other, or both. Flying almost daily in small airplanes above the windswept Patagonia landscape was the most obvious risk. Doug's lifelong aversion to seatbelts ("they give you a false sense of security") while speeding 80 mph on rural dirt roads was another. But a kayak accident near his home? Despite a grief she described as "a hatchet to the head," Kris sought to channel Doug's strength. "It was not a loss," she said. "It was an amputation."

Doug's carpenter and furniture makers in Patagonia worked all night as they crafted a solid casket from Alerce wood. Known as the "Redwoods of the Andes," the towering Alerce trees had initially motivated Doug to defend the forests of Chile. After a quarter century of nonstop battling, Doug Tompkins lay on the coppery-red swirl of fine-grained Alerce plank. Finally, he was getting a rest.

At the Puerto Varas office of Tompkins Conservation, a memorial service was organized. Charter flights arrived. The Chilean first lady, Cecilia Morel, as well as Chilean politicians and friends from Washington, DC, streamed into the tiny Puerto Montt airport nearby. They passed a sole protester, who greeted the arriving mourners with a sign that read "Patagonia Without Tompkins."

Under a tent erected in the yard of the office, friends and colleagues gathered. The public was invited to pay its final respects. Kris made her way through the crowd, speaking with mourners individually, hugging many. The depth of Tompkins's alliances (and influence) was showcased by the presence of former adversaries. Bernardo Matte, the leader of one of Chile's wealthiest families, paid his respects, even though Tompkins had called him a wolf in sheep's clothing in an infamous magazine advertisement. Matte was distraught. Over the course of years, via endless emails, chats, dinners, and trips, the Matte family had learned to appreciate not only Doug himself but also his passionate defense of nature.

Opening the service, Kris could barely speak. She was utterly wrecked by the loss of her Doug. The mission would continue, she declared. Millions of acres of wilderness, roaring rivers, and untouched glaciers—they were close to saving it all. Their forests, ranches, private parks, and nature sanctuaries were assembled and ready to be given as gifts to the people of Chile and Argentina. They would be donated to the public as fully developed national parks. "She spoke with dignity and with a force that welled from a place inside her," wrote Ridgeway. "She gave everything to each sentence and paragraph. Drained, she paused, breathed, and with each breath would rebuild until she continued with an even more profound power that none of us had seen before."

After the memorial service, Kris and inner circle friends boarded flights to Valley Chacabuco where Doug would be buried. A friend's plane was reconfigured. Seats were removed to provide space for the casket. The pilot, Rodrigo Noriega, had learned bush pilot skills with Doug, so Kris entrusted him to fly her husband's body and spirit back to the valley. Noriega flew toward Mount San Valentin, the highest peak in Patagonia. Capped by three glaciers, the flattened peak was a place where Tompkins often sought solace. He liked to gas up the Husky and spiral around the cloud-shrouded summit—his way to blow off steam.

As Noriega approached, the clouds opened, sunlight glinted off the ice. Kris moved into the front seat. Noriega dipped lower, banked over the glacier, and flew one loop, then another. The light plane zoomed over the summit, above the forests and rivers that Tompkins fought to protect. Kris reached behind her, putting her hand on Doug's casket. Describing Doug's battles to prevent the dam on the Baker River, Noriega said, "There's some romance in the sense that Lake Carrera is where the Baker River is born, where it starts. So, he saved the Baker River, but the lake took his life."

Doug's final flight was met by Summer and Quincey Tompkins. His two daughters had traveled nearly twenty-four hours from San Francisco and rushed to the heart of Patagonia. "He was always so invincible—it would have surprised me less if it had been a car accident or a plane crash," said Quincey. "It was very weird. And yet it didn't take me very long to feel like he lived on his own terms, and he died on his own terms. He was there, doing what he loved with his friends. And he was wearing his khakis and his Brooks Brothers button-down shirt, and kind of ill-prepared for what befell him, but that's who he was."

Kris arranged to bury Doug at the tiny Valley Chacabuco cemetery. The graveyard was at the center of their Patagonia Park, next to Doug's beloved airstrip and just a ten-minute walk down the hill from Butler House, their home on the hill. The twenty-plot cemetery had been built by sheep ranchers generations earlier, then upgraded by Doug, who redesigned the fence, replaced the entry gate, and added a sign with a John Muir quotation: "There is no synonym for God so perfect as beauty."

Patagonia's earlier pioneer families faced the frequent death of small children, and many of the region's older gravestones marked the deaths of infants. Burying the babies on the ranch allowed family members to visit their lost children, and for a spouse it was the same. Kris told Doug's aide Hernan Mladinic that she needed Doug nearby. "Doug lived one hundred lives in one," said Kris. "I hope he's remembered as someone who really sought beauty in his life, whether it was the line skiing down a glacier, to architecture, to fine art, to the parks where we created our homes—everything. A renaissance man. A wild guy. Happy and fearless."

Family, friends, and colleagues spoke at Doug's memorials—and hundreds of letters and messages poured in—but it was his closest friends who had the final word. "We could not even miss one paddle stroke for fear of going over," said Yvon Chouinard, recalling the day of the fatal accident. "The boys made a valiant effort—but we lost Doug. We lost a Chief. By his actions, Doug became the teacher we all needed—and he still is."

As the coffin was lowered down, the pallbearers strained. The ropes were tough to maintain, his coffin twisted sideways. The Alerce wood was so heavy it felt like they couldn't hold it up. Mimicking Doug's heavily accented gringo Spanish, a voice shouted "No. No. It's Crooked. Straighten that out. Straighten that out! Straight!" The entire funeral ceremony cracked up. Laughing and crying, they all smiled.

They all knew that Doug would most certainly have fixated on the angle and aesthetics of a coffin being slightly tilted as it was lowered. That was just one of the many Doug voices they already missed.

As shovelfuls of topsoil hit the lid of the casket, Kris tossed a bouquet of native flowers on the casket, and a lone female voice cried out their beloved battle cry—"*Patagonia sin Represas!*" The gathered tribe responded in a raucous chorus—"Patagonia Without Dams!"

Following the funeral, most of the invited guests flew home, but the Tompkins tribe bunkered down in the guest lodge and in the staff quarters clustered at the heart of Patagonia Park. The five survivors of the expedition (Chouinard, Ridgeway, Ellison, Boyles, and Alvarez) gathered in a circle at the lodge to dissect with Kris the details of Doug's last journey. They reexamined the accident in painful sessions—part group therapy, part closure.

Kris needed details. So did Doug's close friends and Summer and Quincey, his daughters. How had those experienced expeditioners, with a record of an ascent of K2 and dozens of kayak first-descents, been ambushed during the equivalent of what they thought was an afternoon stroll? As she peppered the team with queries, Kris singled out the bravery of Weston Boyles. She wasn't looking to blame, but she simply needed to understand.

Reliving the chaos of the final hectic moments and aided by hindsight, the kayakers found several things they could have done differently—like all of them wearing dry suits. But there was no denying the bad luck. The storm came up sudden and deadly. Weather station data later confirmed that gusts had reached sixty miles per hour, while breaking waves on the lake measured six feet. Kris was still in shock. "The loss of the physical presence of him is really painful for me," she said. "It may always be there for me, I don't know."

She then took over their agenda, turned up the heat, and dedicated her entire being into completing their mutual dream. Month after month she would focus her life on completing their mission. "People ask me,

'How did you hold it together after your beloved husband died?' I say, 'It's the reverse: it's probably what keeps me going,'" she said later. "The pressure cooker really took the skin off my back. It was probably the best thing that could have happened to me."

Less than a week after the funeral, Kris flew to Argentina to join President Mauricio Macri in a private meeting to discuss Iberá National Park. Macri leafed through the coffee table books Kris had brought and marveled at the scenes—flocks of white egrets, rainbow macaw parrots, lazing anteaters roaming the marshlands. He saw something like the African savannah but flush with South American wildlife. Kris explained that Iberá was a biodiversity wonderland, one of the few remaining ecosystems on Earth providing natural habitat to flaming red parrots, long-legged wolves, and stalking jaguars. Macri jumped at the chance to provoke a regional economic boom based on nature-as-product and powered by a sustainable development model. What was the downside? Macri asked his ministers. Who could criticize him for receiving privately owned lands and converting them into a national park for all Argentines? That the donation came from a wealthy California couple had made the deal sweeter. "It was a tectonic shift. And Doug's not here," said Kris. "And maybe it happened *because* he's not here. Certainly, it would be impossible to say that Doug's dying didn't have some role in this taking place. Because there was this outpouring of loss for this man who was so extraordinary in many ways."

Within days of his death, Kris noticed a phenomenon in Chile: now that Doug was no longer alive, nearly everyone appreciated him, even former enemies. Tormenting Tompkins had been a Chilean national pastime, but his sudden passing cleared the air of any doubts. Conspiracy theories melted. The undeniable truth of Doug Tompkins was exposed: he had battled to buy up paradise, then fought like hell to give it all away. "When Doug died, things went into hyperdrive," explained Hernan Mladinic. "The attitude of government officials changed. They saw he was a guy who died doing exactly what he believed in. The bureaucrats blocking us probably thought, 'We're not going to be the ones who prevents this project from happening!' He went from public enemy number one to poster child."

A coalition of four Chilean senators suggested that in honor of his contributions to conservation Tompkins deserved the title of honorary Chilean citizen. "He was a man who contributed quite a bit to the national conscience, and he showed business leaders how to give back much of what they earned," explained senator Juan Pablo Letelier.

The measure passed unanimously. "One would have liked to give him this in life, but nobody expected his death; he was so full of vitality," said senator Alfonso de Urresti, who had shared "many fights and battles" with the pugnacious gringo.

Chilean president Michelle Bachelet was also in shock. She'd spent months pushing her cabinet to find solutions, to work with Doug and Kris on the mega-donation. Bachelet lauded Tompkins as "an innovative man, generous in protecting the natural heritage of the planet. His masterpiece is on a global scale and speaks to a vision of gratitude and commitment to future generations." She agreed to meet Kris Tompkins, and suddenly the Chilean government also viewed the couple's mega-donation as a national priority. "I felt an opportunity with President Bachelet. It was a moment," said Kris.

Kris understood that her husband had been a workaholic, but only after his death did she realize how much of his vision existed not on paper but only in the depths of his memory. As she investigated, she marveled at how an *Encyclopedia Britannica*'s worth of details were stored, sorted, and prioritized in his mind. She also felt free to steer Tompkins Conservation in a brave new direction. Kris was determined to push their historic conservation accords with the governments of both Chile and Argentina, but now she would do it her way. She valued bringing long-term stability to an organization that at times jumped from one brilliant idea to another. "I would say his impact on me, which I consider mighty, was from the day he died forward. The rest was bootcamp," Kris said. "As Doug always said, 'We have to be married forever, because who else could stand being married to us!' And in fact he meant it seriously, and he was right. And then he died, and I'm doing the same stuff I was doing when he was alive. But now, the worst thing that could happen to me happened."

In a bold and decisive first move, Kris favored a strategy that Doug never would have: she asked for help. "There was probably no scenario in which we would have done work for Tompkins Conservation had he not passed away," admitted Jib Ellison, Doug's expedition partner and founder of Blu Skye, the environmental consulting firm. "Kris was an unbelievably grieving widow," Ellison explained. "And in terms of all the little details, Doug was really the only one who knew everything that was going on. In the midst of just having lost the love of her life, she was in no position to interview all the different people who thought they reported to Doug, try to get a handle on all those threads, work with the finance department to pull it all together. Somebody needed to take stock, put some structure outside of what existed in Doug's mind."

Kris and Jib worked closely for weeks, then months, to develop a five-year strategic plan. They prioritized dozens of projects, each with a lofty goal, ranging from Doug's fascination to develop Moritz EIS, the tastiest gourmet ice cream ever, to designs for a book to be titled *How to Make a National Park*. "When he died, one of my main goals was to really look at where we are," Kris said. "Because the way I would run an annual financial budget and the way Doug would run it are completely different."

Relations between Kris and Doug's first wife, Susie, were pleasant and never overly complicated. Rare was the year or two when they didn't touch base. This was different. Doug was dead. For the daughters, especially Summer, there was a distinct feeling that much of what had needed to be shared with her father was never spoken. The ending of Summer and Doug's relationship felt incomplete to her. Then,

Doug's last will and testament exploded the precarious relationship with Summer.

The will codified what Doug had already read into the public record for decades—that he would leave no inheritance to his daughters and that all assets would end up with Kris and Tompkins Conservation. But when his daughter Summer read his will, she was stunned. Doug in effect indicated in his will that he was leaving none of his millions to his daughters and he warned them not to contest the will. For Summer Tompkins it felt like a slap from the grave. "That's dark and cruel," said Susie, who was shocked, given that the families were on fairly good speaking terms when he died. "It wasn't like we had been an estranged ex-family," she said.

Quincey and Summer Tompkins were not wanting for cash. Their mom, Susie, pulled in millions from her share of Esprit. Doug had long proclaimed his belief that leaving an inheritance would strip his children of initiative. He had funded a generous emergency medical fund if it were ever needed, but refused to subsidize their lives if they each remained healthy. But he was always clear that every last dollar of his fortune ought to be spent on nature.

Summer, however, criticized her father in the press in both California and Chile, and then sued his estate in both places. She sought a judicial solution to what ultimately was the psychological hole left behind by a father who had never bonded with her—a dad who went on a six-month trip with friends the day after she was born.

Summer had never understood her father's indifference. Considering how she was in many ways quite like her dad—quick of mind, feisty, and confident—was Summer too much a reflection of himself for Doug Tompkins to really let into his life? Summer often had a contentious relationship with her father. He was impossibly bossy and judgmental, but just before his death she confided to her mother that after decades of distance "he was just starting to accept me." A former girlfriend confessed that Doug was ridiculously bad at showing affection for his daughters, while in private he was more generous in spirit. "He loved both of them a lot," she said. "He would sometimes, in very tender moments with me, he would tell me how much he loved those two, how much he loved his children. I wonder if he ever told them."

Kris Tompkins meanwhile focused on her key priority: donating the parklands. In Chile, Kris was certain that President Bachelet was her best—and perhaps final—shot to complete the park plans in a single mega-donation. The terms of the offer were complicated in the exacting details but simple in overall scope. Tompkins Conservation would donate 1.2 million acres of land with a combined $90 million worth of infrastructure to CONAF—the Chilean National Park administrators. In exchange the government would bundle together ten million acres of federal lands and promise to create five new national parks, expand three others, and launch a new era of economic development for Chilean Patagonia.

Kris worked nonstop with Debbie Ryker, Carolina Morgado, and Ingrid Espinoza. They called in all their allies. Kris also asked Jib Ellison to assemble a Blu Skye team to crunch the numbers and brainstorm a five-year strategic plan for the donation. How should she convert her highly subsidized private parks into long-term and self-sustaining national parks run by the Chilean and Argentine governments?

In Argentina, good news continued to arrive. Unlike the maddening pace in Chile, the Argentine government seemed eager to advance plans for Iberá Park. Despite turf wars between provincial and federal authorities, the Tompkins team of Marisi Lopez, Sofia Heinonen, and Corrientes senator Sergio Flinta proved unstoppable. They also had a legion of local supporters. Empowered by the potential for the park to bring in tourists and much-needed revenue, mayors moved to build museums celebrating the local Guarani culture. Just a decade earlier, the brain drain to urban cities like Buenos Aires was leaving the community without hope, but now the young students were returning home proud. They described Iberá and Corrientes as a paradise being rebuilt to its former glory.

"It's too bad Doug is not here today to see that in Corrientes we have proudly saved an ecosystem, in the middle of the extinction crisis," said senator Sergio Flinta, the once staunch critic of Tompkins. "I don't think everything has to do with rewilding species. Something greater was accomplished here. Doug and Kris managed to integrate human beings into the project. This is not an elitist project done by intellectuals or fanatics. As important as the reintroduction of Corrientes' iconic jaguar is, as a symbol of this project, for me it's just as important to see our youth becoming guides for the tourists."

Chapter 22

Islands in the Storm

To me, Doug's achievement is more than just creating a park. He also managed to get a stubborn bunch of people to fall in love with it. Think about it: he was fighting with almost the entire community to get them to fall in love with nature. That's rare, to say the least. And yet he managed to get us involved with it enough to change our perception about living next to nature. That's very powerful.

—**CLARA LAZCANO**, mayor of Chaiten,
a town near Pumalín Park

The death of Doug forced Kris to work triple time. Although his environmental restoration projects lived on, Doug's sudden death left innumerable unfinished projects yet finite financing possibilities. Which ones were crucial to complete his life's work? Kris divested from Moritz Eis, the ice cream company, and sold to Nicolas Ibanez what was suddenly a hot property, a Doug Tompkins original. The timing was crucial, just as money was desperately needed. Kris also sold Doug's beloved Laguna Blanca, the gorgeous organic farm in Argentina with its curved swaths of golds and yellows. Kris also cashed in the political goodwill that had arrived in sympathy and regrets as Doug's death catapulted the mega-donation plans forward. They were outlining a five-year plan to transfer *all* the parklands to public ownership.

Visitors to Patagonia Park and Pumalín Park arrived first by the hundreds, and later the thousands. They marveled at the aesthetics of these remote lands. Free-flowing rivers sliced through valleys teeming with freshly sprouting grasslands. Backpackers and campers arrived from Brazil, Germany, England, and dozens of other countries. Puma and guanaco populations were soaring. Former poachers, hunters, and ex-loggers guided backpackers and GORE-TEX-clad tourists along nature trails that, like the Esprit showrooms of the 1980s, were designed down to the last detail.

Campgrounds featured hot showers fueled by solar panels, communal picnic huts, open fields for enjoying starry nights, and groves of trees to shelter tents from driving rain. The beveled rubbish bins matched the construction style of the visitors center and lodge. "He must have done fourteen versions of that garbage can," laughed Tom Butler, the writer and environmental historian. "It might just be the world's most beautiful trash can."

Kris and her team understood that once they relinquished administration, the standards of maintenance and construction inside their insanely beautiful parks would undoubtedly slip. The national park service was not going to paint the docks at Iberá "Tompkins Green" or maintain the solar paneled showers at the Pumalín campsites. But Chilean conservation officials, including Environment Minister Pablo Badenier, accepted the conditions required to inherit Doug Tompkins's collection of parks, lodges, campsites, and trails.

Badenier felt certain that the donation was so historic that he could find larger budgets to administer the new parks. Yet even as the Council of Ministers moved to accept the parklands, letters to the editor in Chilean newspapers and comments on social media pleaded for the deal *not* to go through. Citizens wrote messages warning Kris, "Don't donate the lands to the government! They'll ruin it!" Kris compared the process of donating the parklands to sending a teenager off to college. "You just have to let go. It's not yours anymore," she said. "Doug was always very clear that despite all their faults there are no better institutions for long-term conservation than national parks."

Gifting the parks was just one piece of the puzzle. Kris and finance manager Debbie Ryker needed to find a long-term economic model that would allow the parks to become self-financing. Could the concept of "the production of nature" become a catalyst for economic development? Could they repeat the success of Iberá throughout Patagonia? Kris bet her chips on Doug's idea to brand all seventeen parks as the Route of Parks.

Kris knew that few destinations on Earth offered travelers such variety of forests, ice fields, lakes, undeveloped beaches, ancient glaciers, and wild, untouched beauty. In Doug's vision each individual park became part of a larger, more meaningful, more valuable whole. From an economic development point of view, the Route of Parks was a stunning multiplier. Communities along the route attending to the first trickle of tourists were encouraged by the flow of dollars. Under the Route of Parks concept, a tourist at one end was not competition but rather a potential visitor to each park.

Chilean politicians were finally recognizing the potential for Patagonia to be an example of a conservation-based economy. Instead of gold mines or aluminum smelters, Doug's concept and what he outlined as "the other economy" based on "the production of nature" had won a share of mind in Chile's economic and political elite, which he knew consisted of no more than a few hundred power brokers. "Things that Doug did twenty-five years ago, taking on climate change and species extinction—it seemed subversive in those days. Unfortunately, those are now things that we're talking about all the time," said Mladinic. "He was a counterculture figure, but values and culture have changed. What used to be strange and subversive is now common knowledge and common sense."

Despite his death, many of Doug's most ambitious ideas lived on. His commitment to bringing the jaguar back to Iberá wetlands was slowly notching milestones. In March 2017 a male jaguar named Chiqui was shipped from Paraguay to the jaguar center in Iberá. Soon thereafter, Tania, a three-legged female jaguar, survivor of zoos and circus, also arrived. A mating program commenced and little more than a year later, while traveling in Kenya, Kris received a video of what "looked like two little pieces of tar."

Initially, she didn't understand. Why had her staff sent her a video of melted rubber? Then she realized the "tar balls" were moving—ever so slightly. They were two baby jaguars, only a few hours old. "I just went berserk, because then I saw a tongue, a tail. And, oh my God!—it was extremely powerful for all of us. These things become like B12 shots to your bum." As the jaguar cubs grew, Kris couldn't stop watching the videos, and giggled about too much "jaguar porn."

Using Valley Chacabuco as her HQ, Kris ping-ponged between lectures in New York City, tours of national parks in Tanzania, and visits to far-flung national park projects in South America. In a moment of escape from her hectic schedule, she went hiking with Sofia Heinonen, the master-planner with Doug of their Iberá park. Camping in a rugged northern corner of Argentina jungle named "Impenetrable Park," Kris came to a realization. She was not just making national parks. Her fight was to return the wild—the wild grasses, the wild jaguar, and her own wild identity. "You have to rewild yourself," she said. "You have to recognize how far from 'wild' you personally have become. Because you can't get into this mindset unless you realize the distance that you've come away from wildness. I talk about rewilding the human mind."

Her passion for conservation intensified after the Corrientes province approved ceding some provincial lands it held to the nation of Argentina, thus forming Iberá National Park in addition to the provincial park. While the ink was still wet on that agreement, she vowed to double down on the reintroduction of lost or nearly extinct species on bringing back wildlife. How had she not seen it earlier? Hiking and brainstorming with Heinonen, they vowed to change the name of the Tompkins foundation's Argentine affiliate. Conservation Land Trust would now be known as Rewilding Argentina.

Given earlier success reintroducing the giant anteaters, the pampas deer, the collared peccary, and the red-and-green macaw, the jaguar births cemented her faith that Iberá could be restored. So Kris budgeted millions of dollars for her rewilding commitment then stepped up her worldwide travels, giving speeches to raise the money to restore the lost animals to Iberá. Seven jaguars were now living in the wetlands, all in huge pens. And the apex predator had powerful allies fighting to recoup space, first in the culture and then in the grasslands. The jaguars were set to be released within several years, and would be free to roam the marshlands and—thanks to more than a decade of education and much patience—many Correntinos were now passionately rooting for the wild cat.

In January 2018, Kris Tompkins received the final approval from the Chilean government for her "all or nothing" offer of a mega-donation. "It was tense, high stakes poker," said one colleague. The Bachelet government accepted the terms of a historic environmental conservation agreement which created five national parks and in total added ten million acres of new national parklands. A ceremony was arranged on the grounds of Pumalín Park.

On the morning of the official transition, as several hundred guests gathered for the formal handover of the parks, the deal was still not sealed. Kris and Carolina held firm. "All the areas are included or the deal's off," they declared. Chilean president Michelle Bachelet was set to arrive by helicopter within the hour. The press and dignitaries were already assembled, yet her team was sorting a flurry of logistical details. The negotiations had not been sealed—a last-minute debate erupted among her staff about missing terms in the documents. They rushed to prepare and print the final agreements. As Kris finished breakfast with the Tompkins Conservation board members, Hernan

Mladinic rushed in with news. "The figures are off!" he declared. "We just found another 143,000 acres"—then, smiling, he added, "rounding error."

Taking the outdoor podium before a crowd of fellow activists, political allies, and a who's-who of the Chilean government, Kris stumbled over her words. She was nervous and took her hands off the papers, then a breeze scattered her speech across the lawn. Gathering her breath and her notes, Kris looked up: an eagle circled above the ceremony. Kris was struck—she was positive it was Doug soaring above them for an aerial view of the historic proceedings. It was surely a scene he had envisioned many times in the twenty-five years since he had launched his bold conservation plans when, on a whim, he bought the original Reñihue farm.

As media headlines worldwide touted Chile's new national parks as a historic conservation victory, many environmental activists who had battled and chafed with Doug over the years saluted his achievements. Mark Tercek, the CEO of The Nature Conservancy, celebrated his accomplishments. Tercek held terrible memories of his personal interactions with Doug. They had argued and clashed. Doug thought TNC was far too accommodating, too willing to compromise. Leaders of TNC, a hugely successful conservation alliance, viewed Doug as a lone kamikaze pilot willing to crash and burn rather than change course to face reality. But when Chile's new parks were announced, they had no doubt that Doug's wildly optimistic park plans had come about in large part because of what Tercek admiringly characterized as his "single-minded vision, huge determination, indifference to critics, doing it his way." Said Tercek, "I think for conservation to happen at any kind of scale in this complicated modern world, you need some kind of extremist."

"He is in the category of Steve Jobs, Bill Gates, and Henry Ford," said William Ginn, the former executive vice president at The Nature Conservancy. "I have personally spent some $3 billion and saved maybe three million acres over the past twenty-two years. He was far more effective."

Following the park handover ceremony, the Chilean government ordered that official maps of the country now include the five new national parks. And in a posthumous recognition of all that Doug had done for the nation, they officially designated his first park as "Pumalín Douglas Tompkins National Park." Doug would likely have chafed at the self-promotion, but after raining criticism and doubts on the motivations of his bold ideas, the government was belatedly celebrating his defense of nature.

Throughout his life Doug Tompkins surrounded himself with breathtaking art, including paintings, quilts, and architectural masterpieces—be they showrooms or campgrounds. Yet it is this strengthening of seventeen national parks including forests, glaciers, and lakes that is most enduring. The Route of Parks stretches for hundreds of miles down the narrowing tail of South America, a dash of wild so extensive that just as astronauts can make out the Great Wall of China they can also note this swash of green—*Tompkins Green*.

A year after the creation of the new national parks, a video surfaced from Patagonia National Park. In the foreground, wind whips the grasslands behind Doug's tombstone. The footage is shaky and ill-framed, clearly a handheld cell phone video. Then a figure emerges from the golden hues. A mountain lion strolls along the hillside. The camera pans a few feet and stops on a second puma. Then there is more movement—a third, a fourth, a fifth, a sixth, and then a seventh puma gather. The mountain lions laze on the hill, just above Doug's tombstone. Two puma then climb up on the cornerstone pillar of the graveyard. Gazing out like sentinels, the pumas survey the landscape. Kris watched the video over and over. Had the pumas come down from the mountains to visit Doug?

Then a remarkable event occurred in Argentina with the jaguar. After a decade of hard work, the Tompkins conservation team in Iberá completed what was essentially Mission Impossible. Despite the hurdles, they had raised a litter of jaguar cubs with such minimal human contact that they were essentially wild. During one of just two interactions with humans, one cub viciously scratched at the vaccination team trying to administer a dose. The jaguar cub was clearly savage and this pleased Kris to no end.

As the world prepared to go into COVID-19 lockdown, two jaguar cubs and their mother were released on San Alonso island in the heart of Iberá. Another litter was set to follow, providing hope that a baseline jaguar population could be established. "For us, it is incredible . . . seeing these animals leaving the pen in the video and leaving their footprints in the middle of the Iberá wetlands," said Sebastian di Martino, the biologist. "We've had the opportunity to spot them twice already, free in the park."

After nearly a hundred-year absence, the magic of wild jaguars in Iberá returned. For dozens of conservation activists it was the culmination of eight years of bureaucratic battles, scientific investigations, and an intense cultural campaign to clear the trails for the return of the revered *yaguarete*. Magalí Longo, a biologist who worked on the jaguar rewilding program with Tompkins Conservation, defined the breeding and release program as essential restoration. "We're fixing the damage," she said. "And it feels great to start seeing results. We're working to make our jobs extinct, but that's a good thing."

Despite the joy at the creation of the parks, to those who knew Doug Tompkins it was clear he would never have celebrated. Yes, the preservation of millions of acres provided refuge for animals, and, yes, the jaguar and puma were reclaiming their habitat. The enthusiasm of rural communities to tie their economic future to the defense of nature was also a valuable victory. But Doug always knew he was losing the greater struggle. A human population topping seven billion, fueled by a carbon-charged lifestyle, meant that his parks remained biological islands. As the greater storm raged, as forests burned and species fell extinct, a broader revolution was needed.

Doug Tompkins never felt there was any other choice but to play a part in the solution. He'd been warning of the dangers of technology and globalization since the early '80s, when he argued with friend and neighbor Steve Jobs. He'd been lecturing politicians, friends, and business leaders for decades. But he always expected leadership to come from below, from the street, from the people. Always the rebel and a strong believer in action rather than proclamations, Tompkins left a legacy not only in the millions of acres he saved but also the thousands of people he influenced.

In countless ways Doug Tompkins has already seeded his next revolution. Emerging environmentally sound businesses—often run by his former employees—are now sprouting up across Patagonia. Nadine Lehner, his assistant for years, is now leading wilderness treks with her company Chulengo Expeditions. Thomas Kimber, his energetic twenty-something neighbor in Puerto Varas, found a way to convert recycled beach plastics into Karün, a collection of sustainably manufactured sunglasses sold in France and sponsored by *National Geographic*. Francisco and Javier, farmers who worked with Doug, are now promoting organic farming via Huerto Cuatro Estaciones, their innovative business that brings fresh produce to locals all year round. "Across Chile, ideas that Doug planted are sprouting and gaining momentum," said Nadine. "Now, even a long way down the road, I see a lot of this around me. In Patagonia all the young people who worked with him have now started their own independent 'earth-positive' projects."

Doug Tompkins never flinched from his conviction that, in the words of Ed Abbey, "sentiment without action is the ruin of the soul." To the young activists he met while battling to stop the Japanese whale hunt, Tompkins had posed a challenge: "Are you ready to do your part?

Everyone is capable of taking up their position to use their energy, political influence, financial or other resources, and talents of all kinds to be part of a global movement for ecological and cultural health. All will be useful. There is important and meaningful work to be done. To change everything, everyone is needed. All are welcome."

Author's Note

On Tompkins's Voice

Throughout this book the quotations from Doug Tompkins include a wide variety of sources, including my own recorded interviews with Doug, his interviews with fellow journalists, radio shows, TV documentaries, letters, emails, and impossibly faded fax proclamations that his staff labeled "Ramblings from the Southern Listening Post." Personal journals kept by close friends Chris Jones, Rick Ridgeway, and Dick Dorworth were extremely useful in capturing details of key scenes.

On Translations from Spanish

For much of the last two decades of his life, Doug Tompkins lived in South America, and when conversations took place in Spanish, every effort was made to maintain the spirit—not the literal translation—of the phrases. Given the heavy use of slang in rural Chile and Argentina, the quotations and sayings have been translated to maintain their humor and wry double meanings wherever possible, but no doubt quite a bit is *lost in translation* when one is recounting the tales of an Argentine alligator hunter or a calloused Chilean cowboy.

List of Interviews

A special thanks to the dozens of people who took the time to recount their experiences with Doug Tompkins. This is clearly a partial list—there were many more who helped. Apologies to those names inadvertently left out. Notebooks were lost. Conversations blended together.

LORENZO ALVAREZ-ROOS. River guide and expedition leader who was kayaking with Doug on the day of the accident.

ANDRES AZOCAR. Chilean journalist and author of *El Millonario Verde*, an early bio of Doug.

MICHELLE BACHELET. Chilean president who fought to complete the plan by Tompkins to combine private and public lands, adding ten million acres of land to Chile's national park system.

PATRICIO BADINELLA. Art director and creative force working on the Patagonia Without Dams campaign.

DON BANDUCCI. Kayak friend. Founder of Yakima.

RICHARD BANGS. Rafting friend. Author, travel television host.

ERIN "LOUIE" BILLMAN. Tompkins Conservation employee, Wharton Business School grad who helped as strategic consultant following Doug's sudden death.

EDGAR BOYLES. Cinematographer who ski-raced with Tompkins in the 1960s and remained friends with him for more than fifty years. Fellow pilot.

WESTON BOYLES. Edgar's son, who knew "Uncle Doug" from childhood. River activist with a foundation that teaches kayaking skills to children.

TOM BROKAW. Newscaster and author who accompanied Tompkins tracking tigers in Russia. Kept photo of Doug on his desk.

KEN BROWER. Environmental activist and author. Son of David Brower, former president of the Sierra Club, an inspiration for Tompkins.

PETER BUCKLEY. Northern California conservation activist and farmer who was Tompkins's friend, confidant, and business partner who purchased a 200,000-acre rain forest in Chile that he donated to create Corcovado National Park.

TOM BUTLER. Environmental historian and author who wrote, edited, and published dozens of books for Tompkins Conservation. Editor of the journal *Wild Earth.*

FRITJOF CAPRA. Author of *The Turning Point*, Tompkins's friend, and activist collaborator.

SERGIO CARDENAS. Chilean secret agent who investigated Tompkins during 1994–1998 "dirty tricks" campaign. Now working in "parks maintenance" in Chile.

NICOLAS CARRO. Biologist specialized in Jaguars (*yaguarete*).

JOHN CASADO. Designer who created the emblematic Esprit logo, then went on to design for Apple MacIntosh.

JUAN EMILIO CHEYRE. Commander in Chief, Chilean Army, 2002–2006, who broke ranks and got the armed forces behind Tompkins.

YVON CHOUINARD. Inventor, mountain climber, and founder of outdoor gear company Patagonia. Best friend of Tompkins from the age of sixteen.

ALDO CIBIC. Italian architect/designer who worked with Tompkins in the 1980s as part of Milan design movement The Memphis Group.

ENRIQUE CORREA. Chilean human rights and democracy activist, later a consultant who helped Tompkins navigate the nation's political minefields.

ELIZABETH CRUZAT. Advertising designer working on the Patagonia Without Dams campaigns.

BOB CUSHMAN. Head of ski patrol at Squaw Valley and veteran outdoorsman who traveled in a small plane with Doug Tompkins on their epic 1989 trip to South America.

DANIEL DANCER. Photographer, worked with Tompkins on the books *Clearcut* and *Overdevelopment, Overpopulation, Overshoot.*

JOHN DAVIS. Environmental scholar at the forefront of direct action and efforts to save wilderness areas in the US. Key figure in shaping Tompkins's grants to environmental groups.

JUAN RAMON DIAZ. Wildlife photographer. Photographed Iberá book for Tompkins.

RICHARD "DICK" DORWORTH. World's fastest skier in the 1960s, lifelong friend of Tompkins, and member of the 1968 "Fun Hogs" group that climbed Mount Fitz Roy.

ALAN DRENGSON. Environmental scholar, author *The Wild Way*.

DUNCAN DWELLE. First-ever employee at Tompkins's 1964 startup, The North Face. Described by colleagues as "Tompkins's Wozniak."

JIB ELLISON. Founder and CEO of Blu Skye, a sustainability consulting firm. Explorer and rafter who organized expeditions with Tompkins on five continents.

MARCI (RUDOLPH) ELLISON. Esprit decorator/merchandiser who later accompanied Tompkins to remote Patagonia in the early 1990s.

STONE ERMENTROUT. Grade school classmate of Tompkins's at Indian Mountain School, fellow skier.

INGRID ESPINOZA. Cartographer and key Tompkins Conservation aide, lived at Reñihue for years. Described as "mad genius."

BILL EVANS. Executive who managed Esprit's computer systems. Saved company by storing hard drives in fireproof safe.

MELINDA EVANS. Esprit design coordinator; oversaw production of clothing lines.

MICHAEL FAY. *National Geographic* explorer, pilot, and conservationist who spent weeks flying with Tompkins in Africa, and in South America.

SERGIO FLINTA. Argentine senator from the Corrientes region. Opposed Tompkins at first, later became his best ally.

DAVE FOREMAN. Daring environmental activist who fought for forests. Founder of both Earth First! and The Rewilding Institute.

SIR NORMAN FOSTER. Iconic British architect tasked by Tompkins to design Esprit's London flagship store in the early 1980s.

JÜRGEN FRIEDRICH. Swiss clothing manufacturer, strategic partner in Esprit International. Later donated millions of dollars to conservation campaigns in South America.

FABIAN GABELLI. Animal trainer for film who then trained parrots to return to the wild in the Iberá wetlands.

WILLIAM "BILL" GINN. Former executive vice president of Global Conservation Initiatives, The Nature Conservancy.

HAROLD GLASSER. Professor of environment and sustainability, hired by Tompkins to run Foundation for Deep Ecology.

DANIEL GONZALEZ. Chilean forestry engineer hired by Tompkins to organize his land acquisition plans. Spent weeks on horseback in the outback scouting lands with him.

HENRY GRUCHACZ. General manager, Esprit clothing company.

PEDRO PABLO GUTIERREZ. Tompkins's longtime attorney. Battled various government efforts to expel him from Chile.

PETER HARTMANN. Chilean environmental activist who worked with Tompkins on the *Patagonia sin Represas* campaign. Founder CODEFF.

RANDY HAYES. Kayak friend. Environmental activist. Founder, Rainforest Action Network. Director, Foundation Earth.

SOFIA HEINONEN. Argentine wildlife biologist and executive director of Rewilding Argentina

BART HENDERSON. River guide, Biobío.

MAL HOLLAND. Next-generation marine activist inspired by Tompkins.

NEWSOME HOLMES. Veteran river-rafting guide who led Tompkins down Zambezi River in Zambia.

DAN IMHOFF. Author, musician, and Esprit employee who became Tompkins's son-in-law.

CATHERINE INGRAM. Friend, partner, and author of books on dharma. Introduced Doug Tompkins to the Dalai Lama.

IGNACIO JIMENEZ. Spanish wildlife biologist who spent a decade with Tompkins preparing to reintroduce anteaters and jaguars to the Argentina grasslands.

CHRIS JONES. Climbing historian/author who climbed with Tompkins on successful ascent of Mount Fitz Roy during 1968 "Fun Hogs" expedition.

CATHERINE KANE. Esprit employee, went on Biobío raft trip.

GERDA KAINZ. Master seamstress, clothing designer. Worked at Esprit.

ROBERT KENNEDY JR. Environmental attorney and avid kayaker who joined Tompkins on far-flung expeditions.

BILLY KIDD. Olympic and world champion skier who trained and raced with Tompkins in the early 1960s.

THOMAS KIMBER. Tompkins's neighbor in Puerto Varas. Founder of Karun.

ANDY KIMBRELL. Writer, editor of *Fatal Harvest*, member International Forum on Globalization. Executive director of the Center for Food Safety.

RICK KLEIN. Forestry engineer and national park ranger who worked twenty-plus years in Chile. Brought Tompkins to the Alerce forest, and Reñihue for the first time.

STEVE KOMITO. Second-ever employee hired by Tompkins to work at The North Face.

LEONARD KOREN. Artist, author of books on Japanese aesthetics.

RICARDO LAGOS. Chilean president (2000–2006) who negotiated half-million-acre donation by Tompkins and quashed Chilean military opposition to his conservation plans.

REG LAKE. Pioneering kayaker who completed first descent of three rivers in California with Tompkins, later known as "The Triple Crown."

JIMMY LANGMAN. Publisher, *Patagon Journal.*

CLARA LAZCANO. Mayor of Chaiten, town near Pumalín.

NADINE LEHNER. Executive director of Conservacion Patagonia. Fencing partner with Tompkins on rainy days. Founder, Chulengo Expeditions.

GARY LEMMER. River rafting guide, Biobío River protector.

ROB LESSER. Renowned kayak photographer; accompanied Tompkins on expeditions in North America and Norway.

AMORY LOVINS. Writer, scientist, energy strategist, Founder RMI.

ALEJANDRO MAINO. Helicopter pilot who rescued Doug Tompkins.

JERRY MANDER. San Francisco activist who taught Tompkins the secrets of rebel advertising. Author of *In the Absence of the Sacred*, a book that deeply influenced Tompkins's activism.

JAMES Q. MARTIN. Photographer and filmmaker, worked with Tompkins Conservation.

CARLOS MARTINEZ. Chilean university professor and anti-environmental activist who infiltrated Tompkins's offices, copying documents and allegedly passing them to government officials.

CAROLYN MCCARTHY. Global communications head for Tompkins Conservation. Writer specializing in conservation, travel, and the outdoors.

JOE MCKEOWN. Climbing partner of Tompkins's in 1960s, lifelong friend.

VICTOR MENOTTI. Conservationist, participated in eco-summit hosted by Tompkins.

HERNAN MLADINIC. Sociologist hired by Tompkins to build inroads into highest level of Chilean government.

GEORGE MONBIOT. British writer, columnist in *The Guardian*, conservationist, activist; exchanged ideas on rewilding with Tompkins.

TOM MONCHO. Esprit executive put in charge of the company when Tompkins left on monthslong expeditions.

FRANCISCO MORANDÉ. Architect, lived and worked at Reñihue for years.

CAROLINA MORGADO. Executive assistant to Tompkins for twenty-five years; environmental activist who began by defending the crown jewel of Chile's wild rivers—the Biobío. Executive director, Tompkins Conservation, Chile.

HECTOR MUÑOZ. Chief of Staff to Chilean Undersecretary of the Interior Belisario Velasco.

RODRIGO NORIEGA. Pilot who worked for Tompkins Conservation.

JULIE OGAWA. Esprit employee, went on Biobío River trip.

JUAN PABLO ORREGO. Chilean environmental activist who worked with Tompkins for twenty-plus years. Founder EcoSistemas, a Chilean NGO.

FRED PADULA. Filmmaker best known for Yosemite climbing doc *El Capitan.*

SEBASTIAN PIÑERA. Chilean president who showed support for Tompkins by purchasing and creating Tantauco, his own 300,000-acre park.

PHILIPPE REUTER. French climber, skier, and adventurer who risked his life in a helicopter trying to save Tompkins.

RICK RIDGEWAY. Documentary filmmaker. Travel partner with Tompkins on many climbing and kayak trips. Author, mountaineering books, including *Seven Summits*.

SHARON RISEDORPH. Architecture photographer, worked with Esprit.

BERNARDO RIQUELME. Producer, radio host Radio Chaiten.

TAMARA ROBBINS. Daughter of Royal Robbins, a leading climber and kayaker who traveled with Tompkins. Rafted the Biobío.

HELIE ROBERTSON. Designer who spent the 1980s organizing Esprit catalog and photo shoots. Helped write the book *Esprit: The Making of an Image*.

PATRICIO RODRIGO. Environmental activist and founder of Chile Ambiente, an NGO that worked with Tompkins for years.

EDWARD ROJAS. Architect from Chiloe, worked with Tompkins.

PAUL RYAN. Photographer, filmmaker, friend.

JOHN RYLE. Writer and anthropologist who wrote an *Outside* magazine article about Tompkins.

CRISTIAN SAUCEDO. Veterinarian working with Tompkins in Chile. Instrumental in rewilding success in Chile.

ALLEN SCHWARTZ. High-flying clothing salesman who joined Plain Jane dress company in early 1970s. Fought bitterly with Tompkins over separation package.

VANDANA SHIVA. Indian agriculture activist and collaborator with Kris and Doug Tompkins. Longtime ally in effort to warn of the dangers of globalization.

JULIE SILBER. Quilt expert, curator, Esprit quilt collection.

DAVE SHORE. Veteran rafting guide who descended the Biobío, among many other rivers.

JIM SLADE. Experienced raft guide who descended the Biobío River.

APRIL STARKE. Worked at Plain Jane and Esprit.

DEYAN SUDJIC. London-based writer/broadcaster. Director of London Design Museum.

CLAUDE SUHL. Climber friend.

JIM SWEENEY. A math whiz hired as furniture craftsman by Tompkins in early 1970s to custom-build Esprit employee furnishings.

JANE TAYLOR. US Navy veteran working to save marine mammals with the Sea Shepherds.

LITO TEJADA-FLORES. Author/photographer who traveled with Tompkins and worked for him in Europe and Asia. Original member of the "Fun Hogs" team that summitted Mount Fitz Roy in 1968.

MARK TERCEK. Former Goldman Sachs partner working as CEO of The Nature Conservancy.

KRISTINE MCDIVITT TOMPKINS. President, Tompkins Conservation; former CEO of Patagonia clothing company. In 2018, was named UN Environment Patron of Protected Areas.

QUINCEY TOMPKINS. Eldest daughter who worked at Foundation for Deep Ecology, the nonprofit that her father created to fund thousands of small environmental initiatives.

SUMMER TOMPKINS. Youngest daughter who often faced her father head on.

SUSIE (RUSSELL) TOMPKINS BUELL. Cofounder of Esprit clothing company who was married to Tompkins for nearly thirty years. Progressive activist working in the Bay Area.

OLIVIERO TOSCANI. Italian fashion photographer who worked with Tompkins as they brought Esprit to international fame with their genre-busting "real people" campaigns.

PABLO VALENZUELA. Wildlife and landscape photographer. Barely survived harrowing flight with Tompkins in windstorm.

WIETSE VAN DER WERF. Founder of Sea Ranger Service; Dutch activist inspired by listening to talks by Tompkins on the Sea Shepherd mission as they battled Japanese whaling fleet.

LUKE VAN HORN. Communications engineer. Sea Shepherd crew.

MARIO VARGAS LLOSA. Peruvian writer, politician, and journalist who interviewed Tompkins at his home in the remote forests of southern Chile.

BELISARIO VELASCO. Former assistant minister of interior who clashed with Tompkins for years.

RODRIGO VILLABLANCA. Tompkins's key collaborator in Pumalín Park.

CARLOS VILLALOBOS. Park ranger for Tompkins Conservation.

MATZAL VUKIC. Architect, friend.

LINDE WAIDHOFER. Landscape photographer who worked years in aerial photography with Tompkins.

PAUL WATSON. Captain of Sea Shepherd Conservation Society, the anti-whaling navy that upends Japanese whale hunts and confronts ocean poachers.

ALAN WEEDEN. President of the Weeden Foundation, longtime Tompkins friend and collaborator in the creation of parks.

DON WEEDEN. Founder of the Weeden Foundation, longtime Tompkins friend and collaborator in the creation of parks.

JOHN WHITMAN. Lawyer from Portland, Maine. An eighth-grade classmate of Tompkins at Indian Mountain School.

TAMOTSU YAGI. Noted Japanese designer who worked with Tompkins at Esprit, then went on to Apple and Steve Jobs.

Suggested Reading

Doug Tompkins surrounded himself by books. He had extensive collections in his homes and offices in California, Chile, and Argentina. For readers looking to delve further into the ideas that inspired him, this list cites key books that I used in my background research into his life, as well as significant books from his library and others written by his friends and associates. By no means is this a comprehensive list—that would be another fifty pages. This is just an appetizer, a small taste of what fed Doug's voracious mind.

Wind, Sand and Stars—Antoine de Saint-Exupéry

A Sand County Almanac—Aldo Leopold

Let the Mountains Talk , Let the Rivers Run—David Brower

Conquistadors of the Useless—Lionel Terray

Climbing Ice—Yvon Chouinard

The Technological Society—Jacques Ellul

Esprit: The Comprehensive Design Principle—Doug Tompkins

In the Absence of the Sacred—Jerry Mander

The Last Place on Earth—J. Michael Fay

The Turning Point—Fritjof Capra

The Global Village—Marshall McLuhan

Deep Ecology: Living as if Nature Mattered—Bill Devall and George Sessions

The Arrogance of Humanism—David Ehrenfeld

In the Shadow of Man—Jane Goodall

Fatal Harvest—edited by Andrew Kimbrell

Let My People Go Surfing—Yvon Chouinard

Desert Solitaire—Edward Abbey

Biopiracy—Vandana Shiva

The End of Nature—Bill McKibben

La Producción de la Naturaleza—Ignacio Jimenez Pérez

Wildlands Philanthropy—Tom Butler

The Ecology of Wisdom—Arne Næss

What Artists Do—Leonard Koren

The Monkey Wrench Gang—Edward Abbey

More than 50 Years of Magnificent Failures—Oliviero Toscani

The Uninhabitable Earth—David Wallace-Wells

Encounters with the Archdruid—John McPhee

The Tiger: A True Story of Vengeance and Survival—John Vaillant

In the Footsteps of Gandhi—Catherine Ingram

Sea Shepherd: My Fight for Whales and Seals—Paul Watson

The Sea Around Us—Rachel Carson

Silent Spring—Rachel Carson

The World Is Blue: How Our Fate and the Oceans Are One—Sylvia A. Earle

Flying South—Barbara Cushman Rowell

Wild Thoughts from Wild Places—David Quammen

Rewilding the World—Caroline Fraser

Feral—George Monbiot

Confessions of an Eco-Warrior—Dave Foreman

In the Presence of Grizzlies—Doug and Andrea Peacock

The Consumer Society—Jean Baudrillard

Amusing Ourselves to Death—Neil Postman

Sapiens: A Brief History of Humankind—Yuval Noah Harari

Ecodefense: A Field Guide to Monkeywrenching—Dave Foreman and Bill Haywood

Spillover—David Quammen

Microcosmos—Lynn Margulis

On Beauty—Tom Butler

Some Stories: Lessons from the Edge of Business and Sport—Yvon Chouinard

Overdevelopment, Overpopulation, Overshoot—edited by Tom Butler

Resurgence of the Real—Charlene Spretnak

Abundant Earth—Eileen Crist

Acknowledgments

Doug Tompkins was a rare jewel, his facets brilliant and cut sharply. There is no way to capture all his wild and rich life. But during the four years that I wrote and researched, literally hundreds of people assisted. Kris Tompkins shared with me an honest portrait of her husband and conservation partner. To Yvon Chouinard, I would like to acknowledge that despite his insanely busy schedule, he never failed to answer my questions and read early drafts of this manuscript. To Doug's daughter Quincey Tompkins and her husband Dan Imhoff, I offer gratitude for sharing a memorable afternoon in your beautiful home as we sought to untangle the web that was Doug. To Susie Tompkins Buell and Summer Tompkins, for honesty in describing the difficult sides of Doug. I appreciate your frank efforts to help me understand Doug.

Mike Faye, thanks for the amazing interviews and insider account of your adventures with Doug. Early versions of this book were read and improved thanks to input from Chris Jones, Rob Lesser, Nadine Lehner, Jib Ellison, Rick Ridgeway, Laura Fernández, Edgar Boyles, Linde Waidhofer, Weston Boyles, Lito Tejada-Flores, Dick Dorworth, Hernan Mladinic, Ignacio Jimenez, and many others.

My longtime agent in London, Annabel Murillo of Peters Fraser & Dunlap, was instrumental in helping me hone the original idea for this book. In New York City, George Lucas of Inkwell Management brought the proposal into reality through his tremendous job presenting my ideas to New York's finest. Of all the editors who worked on this, Miles Doyle of HarperOne was a key early supporter as he shaped the first drafts. Sydney Rogers kindly pushed me when I kept asking for more time to rewrite and rewrite. HarperOne president Judith Curr has quietly shaped and helped my career as a writer far more than she realizes! Thanks Judith! Copy editor Mark Woodworth surprised the hell out of me. I expected a regular edit and felt like I'd won the lottery when I saw how much better he made the prose. Chris Waarlo and Jan Holst, the mapmakers at EMK in Holland, were consummate professionals as we jointly pieced together the varied maps.

My gratitude to Doug's personal lawyer Pedro Pablo Gutierrez; he helped me decipher not only Doug but the Chile in which he landed in the early 1990s. Tom Butler was a key source of information and inspiration. Erin "Louie" Billman fought for this book at just the moment when it seemed the entire project was sinking; her tenacious belief in *A Wild Idea* was instrumental. I am also grateful that Sofia Heinonen, who worked so diligently with Doug for decades (from her base in Argentina), was able to make the time to receive me and guide me through his and Kris's Argentine operations. Joyce Ybarra generously shared material for this book from her son Michael Ybarra's interviews with Doug Tompkins. In southern Chile, Carolina Morgado, Pia Moya, and Weston Boyles were spectacular in their understanding of the region and Doug's role in shaping its destiny. Carlos Cuevas and Dagoberto Gúzman were among Doug's most valued colleagues. He spent years with each of them and deeply trusted their analysis of Chile as he navigated through battle after battle.

Far too many people in Chilean Patagonia and Iberá (Argentina) helped me to name them all. But from drinking mate in the rewilding shacks to a BBQ at Rodrigo Villablanca's home inside Pumalín Park, I felt welcomed by the larger crew of environmental activists who inspired Doug and those who collaborated with him.

And finally, to my seven daughters Francisca, Susan, Maciel, Kimberly, Amy, Zoe, and Akira, I expect that you will always have a little Doug Tompkins in you. While your dad was escaping to the third floor to write, I loved hearing your voices and am long accustomed to writing with the seven of you stopping by to check in with Dad and bringing me a beer, a coffee, or a smile.

My dear Toty, you were so patient on this long, long journey. And *wow* did we have fun exploring the parks of Doug Tompkins together. I needed you badly during the long stretches of this book and you pushed me with just the right touch of inspiration!

Last and certainly not least, I want to acknowledge the dedication and help of my researcher and development editor, Bud Theisen. His story-crafting skills, interviews, thorough research, and edits of this book made it all possible. During four years he followed my wildest ideas as we tried to track the path left behind by a most remarkable man—Doug Tompkins. You made this great, Bud!

Jonathan Franklin
Punta de Lobos Chile

About the Author

Jonathan Franklin is an author and investigative reporter based in Santiago, Chile, and New York City. He has been writing for twenty-eight years for *The Washington Post*, *New York Times*, *The Guardian*, and *der Spiegel*. His previous two books have been optioned for film, and his previous book, *438 Days*, was the number two bestseller on Amazon worldwide.

With hundreds of articles, experience in documentary productions, and broadcast experience with *60 Minutes*, CNN, and Voice of America, Franklin has been covering Latin America for more than two decades. His investigation of gold traffickers was portrayed in the series *Dirty Money*, broadcast worldwide on Amazon in January 2020.

JonathanFranklin.com